Prominence in a
Pitch Language

Prominence in a Pitch Language

The Production and Perception of Japanese

Shinobu Mizuguchi and Koichi Tateishi

LEXINGTON BOOKS

Lanham • *Boulder* • *New York* • *London*

Published by Lexington Books
An imprint of The Rowman & Littlefield Publishing Group, Inc.
4501 Forbes Boulevard, Suite 200, Lanham, Maryland 20706
www.rowman.com

86-90 Paul Street, London EC2A 4NE

British Library Cataloguing in Publication Information Available

Library of Congress Cataloging-in-Publication Data

Names: Mizuguchi, Shinobu, author. | Tateishi, Koichi, 1960- author.
Title: Prominence in a pitch language : the production and perception of Japanese / Shinobu Mizuguchi and Koichi Tateishi.
Description: Lanham : Lexington Books, [2023] | Includes bibliographical references and index. | Summary: "Using production, perception, and processing experiments on understudied sentence- and utterance-levels, the authors explore how Japanese prominence is marked, perceived, and processed. The authors argue that Japanese functions as a pitch language, which marks prominence compositionally by lexical F0 boost and phrasal and boundary pitch movement"— Provided by publisher.
Identifiers: LCCN 2023006983 (print) | LCCN 2023006984 (ebook) | ISBN 9781793645852 (cloth) | ISBN 9781793645869 (epub) Subjects: LCSH: Japanese language—Emphasis. | Japanese language—Versification. | Speech perception.
Classification: LCC PL629.E47 M59 2023 (print) | LCC PL629.E47 (ebook) | DDC 495.61/5—dc23/eng/20230519
LC record available at https://lccn.loc.gov/2023006983
LC ebook record available at https://lccn.loc.gov/2023006984

Contents

Acknowledgments

We started this work when we decided to inquire why Japanese English learners are poor at learning English prosody. We first thought Japanese learners were not good at perceiving it well, especially prominence in English. We did not know how to prove our speculation when one of the authors met Jennifer Cole, who was a member of the prosody workshop at the LSA Linguistic Institute held at the University of Colorado in the summer of 2011. She suggested we should conduct a perception experiment based on Rapid Prosody Transcription (RPT), and we did. She also offered to share her experiment materials and results on English participants with us. Tim Mahrt, her Ph.D. student at that time, supported us with our experiments up to the present. Without their encouragement and help for so many years, we could not have completed this book.

In the course of our study, we came to notice that we do not know much about prosody in Japanese spontaneous speech. To the authors' knowledge, no previous work on Japanese has been conducted in RPT and only a few pieces of work have been done on the perception of it. We thank Gabor Pintér, Yukiko Nota, Takayuki Kagomiya, and Tim Mahrt for coordinating a series of perception experiments in our project, in particular during the period of the COVID-19 pandemic after 2020.

We have tried to account for Japanese prosody synthetically. We are worried that future readers of this book might be bewildered to find many interface aspects involved in our study, but we believe we need to discuss these areas to account for prosody. We thank our mentors Lisa Selkirk, Angelika Kratzer, Masatake Muraki, and Shosuke Haraguchi who taught us phonology, phonetics, syntax, and semantics, and have guided us theoretically. We also thank the anonymous reviewer of the earlier draft of this book to give us important comments and suggestions to revise the draft.

It was a welcome surprise that Jana Hodges-Kluck of Lexington Books got interested in our work and encouraged us to submit a prospectus for publication. We also thank Sydney Wedbush, Ryan Dradzynski, Alexandra Rallo, Megan Murray, and Monica Sukumar of Lexington Books, Rowman, and Deanta Global for helping us publish this book.

Our work has been partially supported by the grants of Japan Society for Promotion of Science (#20520440, #24520542, #15K02480), of Kobe University, Kobe College, and the National Institute of the Japanese Language and Linguistics (NINJAL).

Last but not least, we would like to express our gratitude to the participants of our experiments from the bottom of our hearts.

List of Figures and Tables

FIGURES

TABLES

Chapter 1

What Is Prominence?

How Is It Perceived?

1.1 PREVIOUS STUDIES ON JAPANESE PROSODY

Prosody plays an important role in determining the meaning of the utterance (cf. Bolinger, 1961a, 1961b, 1986, among others), but not all parts of the intonational melody are equally relevant. Some parts of utterances are more prominent than others (cf. Halliday, 1967–1968; Jackendoff, 1972; Ladd, 1996, among others). Prominence is a perceptual feature, loosely defined as parts of utterances that are highlighted by speakers as being important through prosodic, syntactic, and semantic cues (cf. Büring, 2016). Prominence is supposed to stand out to listeners, but what determines perceptual prominence in speech is not well understood. Languages vary in how to mark prominence; it is realized segmentally, tonally, syntactically, or not marked at all (cf. Kratzer and Selkirk (K&S), 2020).

This book discusses the prominence of Japanese. Japanese is an agglutinative pitch language whose basic word order is SOV. To the authors' knowledge, no synthetic work has been done on Japanese prosody that accounted for how Japanese prominence works to determine the meaning of an utterance semantically, pragmatically, and informationally.

Japanese literature has a long history of studies on lexical accents and prosodic phrasing (McCawley, 1968; Poser, 1984; Pierrehumbert and Beckman (P& B), 1988; Selkirk and Tateishi (S&T), 1988; Kubozono, 1988, 2007; Ito and Mester, 2012, among many others), but the prosody above the φ-phrase level is understudied.

(1)   Prosodic Hierarchy (Féry, 2017: 36)

υ	Utterance	(corresponds roughly to a paragraph or more)
ι-phrase	intonation phrase	(corresponds roughly to a clause)
φ-phrase	prosodic phrase	(corresponds roughly to a syntactic phrase)
ω-word	prosodic word	(corresponds roughly to a grammatical word)
F	Foot	(metrical unit)
σ	syllable	(strings of segments)
μ	Mora	(unit of syllable weight)

Unlike stressed languages like English, Japanese assigns its lexical accents by bi-tonal HL. Japanese accent is lexically determined and its placement is often unpredictable. Tokyo Japanese is a moraic language and its accent is realized on a mora. It is generally accepted in Japanese literature that the lexical tones play some part in a larger organization of phrases or intonation patterns, but no standard view for phrasal patterns above the word level is established (cf. P&B 1988: 9). McCawley (1968) posits a level of the Minor Phrase (MinP), the Major Phrase (MP), and the utterance. P&B (1988) assume the Accentual Phrase (AP), the Intermediate Phrase, and the utterance. (X)J_ToBI (cf. Venditti, 1995; Maekawa, 2011a, among others) posits the levels of the AP, the Intonational Phrase (IP), and the utterance. P&B (1988: 16) define the AP as the phrase that has at most one pitch accent and whose periphery is marked with an H at the beginning and an L% at the end. McCawley (1968) and Poser (1984) claim that the domain of "catathesis" is the MP. Catathesis (a.k.a. downstep) is a lowering of F0-contour induced by the H*+L pitch accent. Figure 1.1, for example, shows a sequence of accented (A-) words and unaccented (U-) words; in the left figure of the

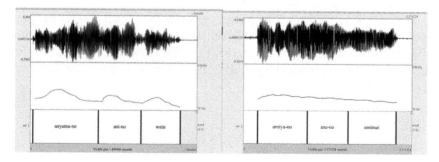

Figure 1.1   Pitch Movement of A-A-A Sequence and UUU Sequence.

A-A-A sequence *A'oyama-no 'ani-no 'wain* "(the) wine of (my) brother in Aoyama,"[1] the F0-peaks of A words gradually lower, but not in the right figure of the U-U-U sequence *Omiya-no ane-no omimai* "(the) sympathy visit by (my) sister in Omiya." The syntactic structure of both sequences is the same, that is, [$_N$, [$_{NP}$ X-no Y]-no W], but we can see that A-words, not U-words, induce catathesis. Tokyo Japanese goes downward as the utterance goes on. The gradual F0-fall in figure 1.1 (right) is not a catathesis but a "declination."

Japanese prosody is complex and is formed compositionally by the lexical pitch accents, phrasal tones, and boundary tones (Féry, 2017: 248). Figure 1.2 shows the pitch contours of the declarative sentence *Naoya-mo oyoida* "Naoya also swam" from Venditti et al. (2008: 488). It is a good example to show that the same utterance is assigned a variety of pitch movements in Japanese.

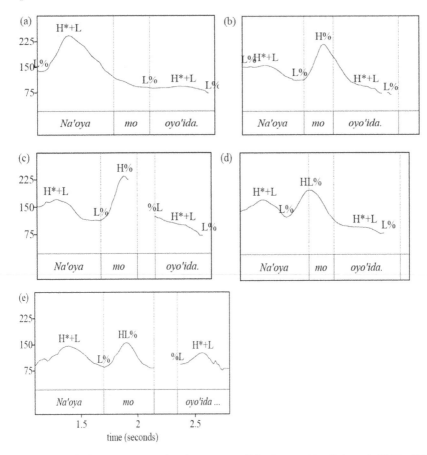

Figure 1.2   Pitch Movements of *Na'oya-mo oyo'ida*. *Source*: Venditti et al. 2008: 488, reprinted with permission from the Oxford University Press.

Venditti et al. (2008) claim that L%, H%, LH%, and LHL% are phrase-final tones and they are assigned at the AP in their framework. Prosodic phrases are very confusing in Japanese literature; Poser (1984) claims that the intonational H% is a final tone inserted at the MP, but P&B (1988: 21) assume that the intonational H% is attached at the utterance level. It is not clear yet at what phrase these tones are aligned. Ito and Mester (2012) claim that there is no need to distinguish an AP from an MP and assume φ and ι phrases that are recursive. Their framework does not assume a phrase that is unique to Japanese like AP or MP but fits in the general prosodic hierarchy (1) in the prosodic literature. To avoid irrelevant confusion in terminology, we will assume the universal prosodic hierarchy (1) in this book.

Prosody carries a variety of roles: phonetic, phonological, syntactic, semantic, pragmatic, and informational functions. The syntax-prosody interface has also been a hot issue in Japanese literature; P&B (1988) proposed that focus appears at the leftmost position of the phrase (i.e., the intermediate phrase in their term) and hence it resets a prosodic phrase. S&T (1988) claimed that Japanese is a left-branching language and promi-nence is aligned leftmost within a φ-phrase. The former is called the "reset theory" and the latter is called the "left edge theory." Both were empirically refuted on the sentence level by the following studies (cf. Shinya, 1999; Kubozono, 2007). But strangely enough, the discussion went on without the definitions of the prosodic phrases, focus, and prominence. Phrases above the word level are not uniquely determined in the literature, as we have reviewed above.

Nespor and Vogel (1986) propose the principle of Relative Prominence, as given in (2).

(2) Relative Prominence (Nespor and Vogel, 1986: 168)
    In languages whose syntactic trees are right-branching, the rightmost node of a φ phrase is labeled s (for "strong").
    In languages whose syntactic trees are left-branching, the leftmost node of a φ phrase is labeled s. All other nodes are labeled w (for "weak").

In a right-branching language like English, the rightmost word of a syntactic phrase is assigned prominence. This is the prominence that is syntactically aligned and is called "alignment prominence/non-focal prominence" (cf. Ishihara et al., 2018). We assume that it is near equivalent to a "broad" focus in the traditional intonational literature, as we will see in more detail in chapter 2.

Focal prominence is also hard to define. It was Jackendoff 1972 who com-bines "focus" with 'stress' and he generalizes that in a stress language like English, intonational pitch accents mark phrasal "focal prominence," and the accents may be high (H*), low (L*), or complex. This type of prominence is generally called a "narrow" focus in the literature. But the incorporation of focus into grammar has ignited a hot dispute in the literature. First, the definition of "focus" is not clear; Wagner (2020) lists a variety of focus types proposed in the literature: "narrow vs. broad" focus, "identification vs. presentational" focus, "contrastive" focus, and "corrective" focus, and K&S (2020) distinguish "FoCus (FoC)," that is, near equivalent to "contrastive" focus, from "information new" focus. Second, the relation between focus and prominence is far from clear. Cross-linguistically, the focus is marked syntactically, morphologically, and prosodically. For example, syntactic markers in English are cleft sentences and *there* constructions (cf. Milsark, 1979). In Hungarian and Spanish, the focus position is syntactically determined (cf. Kiss, 1998; Zubizarreta, 1998, among others). Japanese and Korean have morphological topic/focus markers (cf. Kuroda, 1965; Choi, 1997; Heycock, 2008, among many others). Prosodically, English marks focus on a higher pitch with a lengthened vowel (cf. Liberman, 1975; Pierrehumbert, 1980, among many others). Jackendoff's generalization requires that the main prominence within the sentence must be within the F(ocus)-marked constituent. But as the literature (cf. Zubizarreta, 1998; Kadmon and Sevi, 2011, among many others) correctly pointed out, the semantic focus domain and the accent domain do not necessarily match. "Given/New" are Information Structure notions and they also show a complicated relationship with prosody. Givenness forces the lack of prominence on content words under certain discourse conditions in Standard American and British English (for exceptions, see Schwarzschild, 1999, among others). K&S (2020) point out that contrastive focus (FoC in their term) is signaled by prosodic prominence, but "information focus" merely presents new information without prominence. They claim that information structure CAN have an impact on prosody but does not HAVE to, and there is no necessary link between prosody and information structure. We need to reconsider the relationship between the semantic/pragmatic/informational aspect of focus and prosody. See Büring (2016), Wagner (2020), and chapter 2 for details.

In Japanese literature, no definition of "focus" is found except in Ishihara (2016) and Ishihara et al. (2018), who define "contrastive" focus after Krifka (2008). Strangely enough, the so-called "focus effects" have been much talked about, without a definition of focus; "focal prominence" is analyzed as being characterized by F0 boost and post-focal compression (cf. Kori, 1989, 2020; Shinya, 1999; Kubozono, 2007; Ishihara, 2016, among others). We know that Japanese prosody is complex (cf. figure 1.2) and it is probably formed compositionally by the lexical pitch accents, phrasal tones, and boundary tones. If we are on the right track in incorporating focal prominence into the prosodic analysis in Japanese, we have to first define what focal prominence is. We also believe we must make clear what prosodic phrases Japanese has. It was

P&B (1988) that incorporated focal prominence with boundary chunking. Since discrepancies between syntactic boundaries and prosodic boundaries have long been disputed (cf. Kubozono, 1988, Ito and Mester, 2012, among others), we need to know what function boundaries load in Japanese prosody.

Another problem of the traditional studies on Japanese prosody is that they mostly consider the production patterns on the word level or the phrase level at most. Only some works are found on the sentence level, but most studies are only descriptive. What is more, less work has been done on perception and processing. Prominence is, as mentioned above, a perceptual feature, loosely defined as parts of utterances that are highlighted by speakers as being important through prosodic, syntactic, and semantic cues (cf. Büring, 2016). A study of prominence is not complete without considering its perceptional strategy.

Note here that the perception of prosody in pitch language requires some caution. In pitch language, pitches gradually lower as the utterance goes on, so listeners have to normalize F0 when they perceive it. Pierrehumbert (1979) shows that when two stressed syllables sound equal in pitch, the second one is lower in F0 due to declination (cf. figure 1.1 (right)). Figure 1.3, for example, illustrates two focal prominences on *chocho* "major key" and *tancho* "minor key" marked by the participants of our perception experiment (cf. chapter 6). We can see that the F0 peak of the second prominent word *tancho* is lower than the one of the first prominent word *chocho*, while they sound equally prominent. Pierrehumbert (1979) claims that F0 is normalized when we perceive it.

Shport (2015) studied whether listeners are sensitive to F0 peak location or the pitch movement of F0-fall after the peak when they judge focal prominence. She conducted perception experiments on Japanese and English

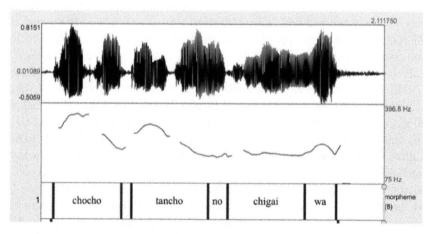

Figure 1.3   Prominences Marked within a Boundary. (extracted from CSJ file #S00f0095)

listeners and found that Japanese listeners are more sensitive to F0-fall than English listeners (cf. chapter 4 for details). She concludes that Japanese listeners have a native language bias toward F0-fall. Tokyo Japanese realizes a "downtrend" as the utterance goes on: it realizes catathesis, declination (cf. figure 1.1), and final lowering. These characteristics might lead to a language bias in perception.

We have pointed out the problems of the previous studies, and we will try to analyze the understudied areas, especially the perception and processing of prosody of spontaneous Japanese in this book. Along with theoretical queries, we have a very naïve doubt on often cited assumption; when we conducted perceptional experiments on corrective focus (see chapter 4 for details), we found that Japanese is not as easy to perceive corrective focus as English. Japanese is our L1 and English is our L2. We believed that L1 was easier to hear than L2. The experiment results were the opposite. We were puzzled and wondered why. So, we conducted a processing experiment for the identification of corrective focus via fMRI and found that the response time is longer to process Japanese corrective focus than to process English (see chapter 5 for details). That was the least expected result. Gandour et al. (2007) conducted an fMRI experiment on late-onset Chinese-English bilinguals for the identification of broad focus and contrastive focus. The task accuracy is higher in L1 than in L2 in their experiment. Though L1 and L2 listeners elicited highly overlapping activations in frontal, temporal, and parietal lobes, they found that greater activation in L2 than in L1 occurred in the bilateral anterior insula and superior frontal sulcus. They conclude that L1 and L2 are medicated by a unitary neural system regardless of the age of acquisition and that additional neural resources may be required to process L2 for unequal bilinguals. We suspect Chinese-English and Japanese-English bilinguals use different substrates.

Another surprise came to us when we conducted perception experiments on spontaneous Japanese and English by native listeners of Japanese (cf. Pintér et al., 2014; Mizuguchi et al., 2017, 2019a, 2020a). Our participants unanimously said that it was much easier to perceive prominence in Japanese than in English. The inter-listener agreement ratio is higher in English than in Japanese, but the participants' impression was the opposite.

We wonder why L1 is not easy to hear. One might say that Japanese prominence would be hard to perceive simply because Japanese did not mark prominence. We do not accept this idea, though. Prominent boundary tones have long been recognized in Japanese prosody literature (cf. Kawakami, 1957; Oishi, 1959; Kuroda, 1965; Kuno, 1973; Tomioka, 2009, among others). Figure 1.2 (a) is different from figures 1.2 (b)–(e) in that particle *mo* "also" is assigned the phrase-final tone L% while the *mo* in figures 1.2 (b)–(e) is assigned H% or HL%. Note that the particle *mo* is a function morpheme with no lexical accent. Boundary tones of H% and HL% mark "focal prominence" on the particle and are analyzed to load a semantic and pragmatic function. These examples alone show that the Japanese boundary tones are responsible for a difference in meaning. We will come back to the semantic and pragmatic role of boundary tones later in chapter 7.

1.2 THE STRUCTURE OF THE BOOK

We would like to investigate how Japanese prominence is marked and perceived. Also, we hope to answer our question of why L1 listeners have difficulties perceiving prominences in Japanese. The structure of this book is as follows.

In chapter 2, we will first review how "focus" and focal prominence have been incorporated into grammar and list the kinds of focus considered in the literature. Then we would study what kinds of focus are marked in Japanese utterances.

In chapter 3, we will conduct production and perception experiments of "information-new" focus on the sentence level and provide quantitative confirmation for the traditional analyses. Japanese focus has been considered to show two "focus effects": F0 boost and post-focal compression. We will show that these effects are not as strong as considered in the literature and F0 is not a strong cue for the perception of focal prominence on the sentence level.

In chapter 4, we will consider corrective focus, based on perception experiments. We compare Japanese with English, showing that focus perception is worse in Japanese than in English by native listeners of Japanese. Our finding is not compatible with the naïve idea that L1 is easier to perceive than L2. We will claim that Japanese is not easy to hear not because it is L1 or L2 but because Japanese focal cues are hard to perceive in particular environments even for native listeners.

In chapter 5, based on the results in chapter 4, we will investigate how corrective focus is processed from a neurocognitive point of view. We will conduct an fMRI experiment on Japanese and English corrective focus by late-onset Japanese-English bilinguals and show the difference in processing between the two languages. Our behavioral data is different from the traditional study (cf. Gandour et al., 2007) in that L1 is not always higher in focus identification rate and shorter in response time. We will show that we use the right hemisphere effectively to process Japanese corrective focus and the processing substrates are different between Japanese and English.

Chapter 6 discusses the perception of prominences and boundaries in the spontaneous Tokyo Japanese. As reviewed in 1.1, the notion of boundaries is not well-established in Japanese literature. We will conduct perception experiments on spontaneous Japanese and investigate how listeners perceive boundaries and prominences on the utterance level. Our method is Rapid Prosody Transcription (RPT) developed by Cole and her colleagues (cf. Cole et al., 2010a). Untrained transcribers mark boundaries and prominences while listening to spontaneous speech, and b(oundary)-scores and p(rominence) scores, that is, the rate of the markings, are measured. Our materials are 12 excerpts from *Corpus of Spontaneous Japanese* released by the National Institute of Japanese Language and Linguistics. We have found that boundaries are cued by syntactic categories. We have also found that phonetic cues of F0, duration, and intensity are not effective prominence cues, though range F0 is a better predictor for prominence in spontaneous Japanese. We recruited three groups of Japanese listeners and found a dialectal perception bias among the dialects we studied. That was an unexpected result that suggests a cross-dialectal perception bias in Japanese.

Chapter 7 concludes our discussion on prominence in Japanese. Our conclusions are that (i) Japanese prominence is focal, (ii) Japanese marks prominence on content words as well as function morphemes, (iii) local F0 boost and boundary pitch movement (BPM) are the cues to mark prominence, (iv) the scope of the focus differs on which cue it is loaded with, (v) BPM is possibly aligned to function morphemes and invokes a pragmatic implicature, and (vi) prosodic boundaries are syntactically controlled. Based on the experimental data, we conclude that prominence cues and processing strategies vary among languages and dialects. Our answer to the question "Why is L1 not easy to hear?" is that some of the cues are hard to process even for native listeners.

We have found that there is a cross-dialectal difference in prominence perception, but we have not accounted for why they show a dialectal difference in perception, yet. There are many dialects in Japanese and they are different in how lexical accents are assigned, how dephrasing works, and which tonal scaling they form. In a dialect with upstep tone scaling, there is a possibility that local F0 boost is not effective for prominence marking. We leave a cross-dialectal and a cross-linguistic difference in prominence perception for future research.

NOTE

1. Japanese does not have definite and indefinite determiners, and its arguments are subject to delete on the surface as far as they are recoverable from the context. We tentatively insert *the* and *my* in the gloss here.

Chapter 2

Non-focal and Focal Prominence

2.1 GENERAL VIEW

Prominence is a perceptual feature, loosely defined as parts of utterances that are highlighted by speakers as being important through prosodic, syntactic, and semantic cues (cf. Büring, 2016). Syntactically marked prominence is called "alignment/non-focal" prominence and semantically marked prominence is called "focal" prominence. Nespor and Vogel (1986: 168) proposed the principle of Relative Prominence as in (1).

(1) Relative Prominence (Nespor and Vogel, 1986)
In languages whose syntactic trees are right-branching, the rightmost node of a φ phrase is labeled s (for "strong"). In languages whose syntactic trees are left-branching, the leftmost node of a φ phrase is labeled s. All other nodes are labeled w (for "weak").

English, for example, is a right-branching language and the rightmost content word of a phrase/sentence is aligned prominence (cf. Wells, 2006). Japanese is a left-branching language and the leftmost word of a phrase/sentence is alignment prominence (cf. Selkirk and Tateishi (S&T), 1988).

The prosody of an utterance plays a significant role in determining the meaning of the utterance. Traditional studies (cf. Bolinger 1961a, 1961b, 1972, 1986; Halliday 1967–1968, among many others) observed that prosodically "focused" constituents play a vital role in processing the utterance. Halliday (1967) classifies the prosodic focus into "broad focus" (BF) and "narrow focus" (NF). For example, (2) is ambiguous in meaning with the prominence of *the shed*. When the prominence is interpreted as BF, the prominence on *the shed* can signal the whole sentence being focused (cf. (3a)). As NF, it can signal the foci on *the shed* (cf. (3c)) or *painted the shed* (cf. (3b)).

(2)   John painted *the shed* yesterday.   (Halliday, 1967: 207)

(3)   a.   [$_F$ John painted the shed yesterday]

     b.   John [$_F$ painted the shed yesterday]

     c.   John painted [$_F$ the shed] yesterday.

It was Jackendoff (1972) who incorporated "focus" into grammar. He defines "focus" as the information in the sentence that is assumed by the speaker not to be shared by him and the hearer (cf. Jackendoff, 1972: 12). He claims that if a phrase P is the "focus" of a sentence S, the highest stress in S will be on the syllable of P that is assigned the highest stress by regular stress rules. In a stress language like English, intonational pitch accents mark phrasal "focal prominence," and the accents may be high (H*), low (L*), or complex.

Jackendoff's generalization requires that the main prominence within the sentence must be within the F(ocus)-marked constituent. In the Question-Answer pairs as in (4), the accented name *Bill* in (4A1) informationally corresponds to the WH-word in (4Q). *Bill* then is analyzed as focus, whereas the accented name *Sue* in (4A2) does not correspond to the WH-word in (4Q), so it is only interpreted as having a mismatched prominence.

(4)   Q: Who did you introduce to Sue?

     A1: I only introduced [$_F$ Bill] to Sue.

        L*   H*L L%

     A2: #I only introduced Bill to Sue.

        L*   H*L L%   (Kadmon and Sevi, 2011: 3)

With the "Jackendovian" focus incorporated into the grammar, it is expected that there is one uniform and the well-defined notion of "focus," with some clear semantic/pragmatic criterion for identifying the focused constituent and highly systematic prosodic marking of the focused constituent.

However, this view faces difficulties in determining prosodic and semantic domains of focus. In English and other Germanic languages, the domain of semantic focus is not always the same as the prosodic domain of focus. Cinque (1993), Zubizarreta (1998), and many others observe that the domain of semantic focus can be larger (cf. (5)), or smaller (cf. (6)) than the accent domain.

(5) Semantic Focus Domain > Accent Domain

    a. Q: What did John do?

       A: [John [$_F$ ate [the PIE]]]

    b. Q: What happened?

       A: [$_F$ John [ate [the PIE]]]   (Zubizarreta, 1998: 2)

(6) Semantic Focus Domain < Accent Domain

    a. We need to [[$_F$ IM]PORT], not [[$_F$ EX]PORT].

    b. Oh, I've only got [[$_F$ THIR]TEEN], not [[$_F$ FOUR]TEEN].

In addition to BF and NF, the notion of "Contrastive Focus (CF)" is much discussed in the literature. CF is the "focus" to describe utterances in which a single word or constituent within a sentence is singled out prosodically (cf. Gussenhoven, 2007).

Rooth (1986) proposes that CF induces a set of alternatives; in Rooth's Alternative Semantics, CF carries a syntactic feature [F], and the H* pitch accent is the prosodic manifestation of this feature. The F feature semantically induces a set of alternatives. *Bill* in (4A1), for instance, carries a syntactic feature [F] and induces the set of propositions that speak of some individual whom the speaker introduced to Sue, for example, {I introduced Bill to Sue, I introduced Tom to Sue, I introduced Jack to Sue, . . .}. *Bill* is chosen as CF and is assigned the pitch accent H*L. Thus, F highlights the contrast and aims for the highest prominence in a sentence.

Although Alternative Semantics has been widely supported and studies on the focus operator *squiggle* ~ (see Rooth, 2016 for details) and its domain are much in progress (cf. Büring 2016; Wagner 2020 for reviews), many researchers cast doubt on this picture. For example, Kadmon and Sevi (2011) ask whether "focus" can be independently identified by prosody, as the connection between focus and prosody is far from straightforward (cf. (5)–(6)). Kadmon and Sevi (2011) argue that a major role of pitch accent placement is to indicate information about "given" and "new." For them, (4A2) is infelicitous simply because *Sue*, which is given, is prosodically prominent, not because it is focused.

"Given/New" are information structure notions and show a complicated relationship with prosody. Kratzer and Selkirk (K&S) (2020) propose a new idea on focus and information structure. They introduce the notion FoCus (FoC, henceforth), which evokes alternatives to a mentioned individual, concept, or proposition, to highlight a contrast. [FoC] needs to be distinguished from information focus ("newness focus"), which applies to expressions that merely present new information. Compare (7) with (8).

(7) Me: Did anybody eat the clementines? I can't find them in the pantry.

   You: (I think) <u>Paula</u> might [have eaten the clementines].

         New, no contrast nor alternative set

(8) Me: Sarah mailed the caramels.

   You: (No), [<u>Eliza</u>]$_{FoC}$ [mailed the caramels].

         FoC, evokes alternatives to Eliza

   (K&S 2020: Examples (1)-(2))

K&S (2020) argue that the grammar of Standard American and British English marks Givenness [G] and FoCus [FoC], but a merely New material ("information focus" in their term) remains unmarked. [FoC] highlights the contrast and aims for the highest prominence in a sentence. [G] is sensitive to discourse givenness and resists phrase-level prominence. There is no representation of newness.

Traditionally, information structure covers concepts related to focus, given-/newness, or topicality. Givenness marking is sensitive to whether an individual, concept, or proposition, has been mentioned before or is present in the context (cf. Chafe, 1970; Schwarzschild, 1999; Rochemont 1986, 2016, among many others). Givenness forces the lack of prominence on content words under certain discourse conditions in Standard American and British English. Jackendovian "focus" requires prosodic prominence, but the definition of "focus," however, has not become clear (cf. Kadmon and Sevi, 2011; Wagner 2020, among others). Wagner (2020) lists a variety of focus types ever proposed in the literature: "narrow vs. broad" focus, "identification vs. presentational" focus, "contrastive" focus, and "corrective" focus, and K&S (2020) further distinguish "FoC" from "information new" focus; FoC is a contrastive focus which induces alternative sets in the sense of Rooth (1986, 2016), and is signaled by prosodic prominence, but information focus merely presents new information without prosodic highlight.

Topicality is another much-discussed information structure notion, and it is often discussed in contrast to "comment," which originated in the Prague School. There is a vague agreement in the literature that the topic is given and the comment is new.

The picture of focus, given-/newness, topic, and comment is thus complex. K&S (2020) even claim that there is no common ground on how the information concepts of focus, givenness, and topicality relate to each other despite many years of research; information structure CAN have an impact on prosody but does not HAVE to, and there is no necessary link between prosody and information structure. If their claim is on the right track, why do we believe that prosody plays a critical role in determining the utterance's meaning? Let us consider where this belief comes from.

An enormous number of "focus" studies have tried to show how prosody is important in information structure with Question-Answer pairs like (9).

(9)   Q.   What did John eat?

     A.   #JOHN ate the pie.

(9A) is infelicitous because *John* expresses "old" information, which is resistant to phrase-level prominence. Schwarzschild (1999), Kadmon and Sevi (2011), Wagner (2020), and many others argue that the story is not this simple, but for the moment, let us assume that the given information is not prominent prosodically.

Another Question-Answer pair familiar in the focus literature is the one like (5b), repeated below.

(5)   b. Q: What happened?

     A: John ate the PIE.

(5bQ) seeks for brand-new information, and the answer of (5bA) is felicitous with the prominence of *the PIE*. Halliday (1967) calls this type of prominence as "broad" focus. The prominence of *the PIE*, however, does not guarantee the newness of the whole sentence. As discussed above in (2), *the PIE, ate the PIE, John ate the PIE* are possible focus candidates. New information does not require prominence, either, as discussed by K&S (2020); as (7) with (8) above show, we can see that it is a contrastive focus, not an informational new focus that requires prominence.

We can account for the prominence of *the PIE* in (5bA) without reference to focus; *the PIE* gets prominence not because it is new information but because it occupies the default position to get prominent in an Intonation Phrase (IP) in English. Recall the principle of Relative Prominence proposed by Nespor and Vogel (1986) in (1). Wells (2006) also observes that English is a language that claims one prominence per IP, and that prominence falls on the primary stress in the last content word of each IP. In (5bA), *the PIE* is alignment/non-focal prominence.

What is confusing with (5bA) is that it can also be a felicitous answer to (9'Q), that is, *the PIE* in (9'A) is focal prominence.

(9')   Q.   What did John eat?

       A.   John ate the PIE.

We assume, following K&S (2020), that *the PIE* in this context is a contrastive focus and evokes an alternative set {John ate the pie, John ate the cake, John ate the pasta. . .}. We have to keep in mind that both types of prominence, that is, non-focal prominence and focal prominence, can be realized with the same sort of pitch movement, as given in (5) and (9').

Languages spell out information structure in various ways: segmentally, prosodically, tonally, or not at all. English demands one prominence per IP due to Nuclear Stress Rule, and prominence is, therefore, aligned to the rightmost content word in an IP by default. It is non-focal prominence. English also realizes contrast by prominence. This type of prominence is focal prominence. Both types of prominence are marked by acoustical cues of maximal F0, duration, and intensity in English. Therefore, judging which prominence prosody marks is sometimes confusing without context.

2.2 A CASE OF JAPANESE

This section reviews how prominence has been analyzed in Japanese literature and discusses how it is marked if any.

Japanese basic syntactic structure is SOV but Japanese allows scrambling, that is, alternation of word order, because it has case particles. Also, arguments are subject to deletion as far as they are recoverable from the contexts, so surface structures often lack necessary arguments in Japanese. Japanese is a left-branching language; the leftmost word of a phrase/sentence is supposed to mark alignment prominence. Pierrehumbert and Beckman (P&B) (1988) claim that the leftmost word in a prosodic phrase marks the Maximal F0 and is the most prominent. In their framework called the "reset theory," a new prosodic phrase is introduced whenever some element is highlighted and F0 is reset to return the F0 to a reference line. Selkirk and Tateishi (S&T) (1988) work on a syntax-prosody interface and propose the "left-edge" theory. They claim, along with P&B, that the prominence occupies the left edge of a prosodic boundary. Both the reset and left-edge theories show that prominence is syntactically aligned to appear at the leftmost element in a prosodic phrase. We can say that the prominence that both the reset theory and the left-edge theory assume is alignment/non-focal prominence.

There has been a heated discussion in the literature on how focus effects interact with catathesis (a.k.a. downstep), and there are two conflicting theories. In one view, focus is strong enough to initiate its major phrase (MaP).

Thus, the existence of focus "resets" catathesis (cf. P & B, 1988) and initiates a new MaP. The other view is that focus simply raises the pitch and lowers the following F0 but that the effects are not phrasal and not strong enough to identify them with phrasal breaks (cf. Shinya, 1999; Ishihara, 2003, 2016; Kubozono, 2007, among many others). The former locates focus at the phrase-initial position only, and the latter allows focus to appear in the middle and the final position of a prosodic phrase. Recently, the latter theory appears to be widely accepted, empirically at least (see chapter 6 for the perception of prominence in spontaneous speech).

The discussion is exciting, but we should not forget one important thing. That is, we could judge whether the two theories or the refutation are correct only when we know what prosodic boundaries are and how they are coded. As we have discussed in chapter 1, prosodic boundaries are vague in the Japanese prosody literature. Also, we do not know how boundaries are perceived and processed in Japanese. We will conduct perception experiments on spontaneous Japanese in chapter 6 and investigate how boundaries and prominences are perceived. We can judge whether Japanese aligns non-focal prominence or not only when we get concrete data.

Let us move on to "focal" prominence. We will review the traditional theories first and then point out problems.

Japanese is an agglutinative pitch language. It has more or less different prosodic properties from stress languages; lexical accents are realized by bitonal pitches H and L (cf. (10) and figure 1.1 for Tokyo Japanese).

(10) a.   Accented:[1]

        Initial-Accented: 'inoti (H*LL) "life," 'ame (H*L) "rain"

        Medial-Accented: ko'koro (LH*L) "heart," u'mai (LH*L) "delicious"

        Final-Accented: ata'ma (LHH*(-L))[2] "head," ma'me (LH* (-L)) "bean"

    b.   Unaccented: amai (LHH) "sweet," ame (LH) "candy"

Japanese also employs pitch to mark focal prominence. P & B (1988) is a pioneering work on Japanese focus, and they observed that the focus prosody of unaccented (U-) words is different from that of accented (A-) words. Figure 2.1 illustrates the pitch movements in Adjective-Noun sequences in which an adjective is focused: (a) the U[+F]-U sequence (*amai ame* "sweet candy"), (b) the U[+F]-A sequence (*amai ma'me* "sweet bean"), (c) the A[+F]-U sequence (*u'mai ame* "delicious candy"), and (d) the A[+F]-A sequence (*u'mai ma'me* "delicious bean"). The initial low %L (L% in P&B's notation in figure 2.1),

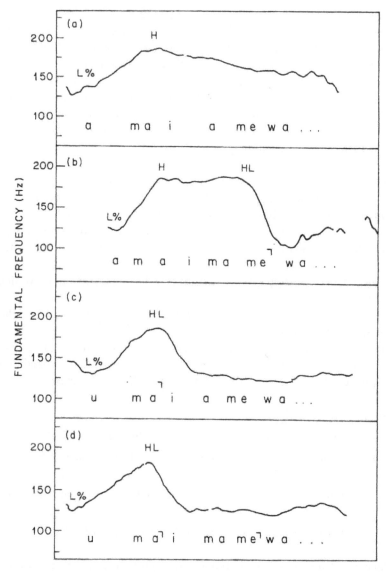

Figure 2.1   F0 Contours of Adj[+Focus]-Noun. *Source*: Pierrehumbert and Beckman 1988: 107, reprinted with permission from the MIT Press.

that is, the boundary tone, is observed in all the pitch movements in figure 2.1 a–d, but F0-rise induced by focus is observed differently between A-words (cf. figure 2.1 c–d) and U-words (cf. figure 2.1 a–b); *u'mai* "delicious" is accented and the F0 peak of H*L is higher when it is focused. *Amai* "sweet" is unaccented and the F0 peak H, which is called phrasal H, is higher when it is focused. In figures 2.1 c–d, the A-word *u'mai* induces catathesis, and the noun *ame/mame* "candy/pea" follow focused [u'mai]$_F$ with a relatively lower pitch. What is interesting is that post-focal compression, a typical characteristic of Japanese focus, is observed only in figures 2.1 c–d. In figures 2.1 a–b, where *amai* "sweet" is unaccented, the phrasal H forms a plateau and post-focal compression is not observed. Figure 2.1 shows that focus effects of F0-rise and post-focal compression emerge only when the A-word *u'mai* "delicious" gets focused.

Though P & B (1988) investigated the differences in focus effects of A-words and U-words, there was not much discussion on U-words after their work. Most studies on Japanese focus deal with Accented (A-) words only, without any reference to Unaccented (U-) words.

It is sometimes difficult even for native listeners to judge what Japanese pitch movement realizes in a spontaneous speech; it could be lexical, phrasal, or utterance information (cf. chapter 1). Moreover, the tonal scaling is gradually lowered in Tokyo Japanese (cf. figure 2.2).

Due to the downtrend, the initial H, whether it is phrasal (-H, cf. figures 2.1 a,b) or accentual (*H+L, cf. figures 2.1 c, d), is usually the Maximal F0 of an utterance. As we will discuss later in chapter 6, Japanese listeners identify no prominence on the initial F0 peak when they hear the utterances in figure 2.2. Instead, some of the listeners, but not all, find the boundary-final *de* (Conjunction) prominent.

Japanese focal prominence is considered to have been well-studied (cf. McCawley, 1968; Poser, 1984; P&B, 1988; S&T, 1988; Kori, 1989; Shinya, 1999; Kubozono, 2007; Ishihara, 2003, 2016; Ishihara et al., 2018; Venditti et al., 2008; Igarashi, 2014, among many others), but perception and

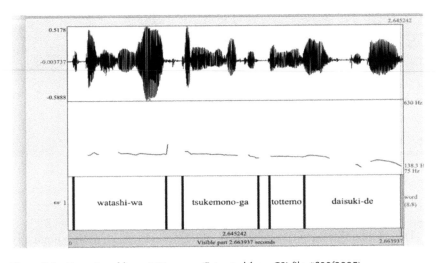

Figure 2.2 Downtrend in an Utterance. (Extracted from CSJ file #S00f0095)

processing of pitch movement are understudied. We need to investigate how prominence is perceived. Otherwise, we cannot see what cues prominence in an utterance.

Also, we need to make clear the notion of "focus"; strangely enough, the notion of "focus" is not clearly defined in most of the previous research on Japanese focus. Ishihara (2016) is a notable exception and restricts his research subject to contrastive focus in his analysis. However, most other studies do not identify which type of focus they are dealing with. As reviewed in section 2.1, many types of focus have been considered in the literature: "narrow vs. broad" focus, "contrastive" focus, "corrective" focus, "information new" focus, and so on (cf. Wagner, 2020). Since the lack of a clear definition of the target of the study is often fatally deficient as scientific investigations, we need to consider what "focus" Japanese realizes by prominence.

In section 2.1, we reviewed the "narrow/broad" focus in English (cf. (2)–(3)). On the assumption that alignment/non-focal prominence marks a broad focus, we do not think Japanese has a broad focus, though further investigation is still necessary before we can conclude. We will conduct a production and perception experiment of narrow/broad focus in chapter 3.

As for the "information new" focus, we can replicate Question-Answer pairs like (5) in English and see whether Japanese marks the information new focus. Also, we have to recall that Japanese is an agglutinative language and is rich in its morphological markings of information structure notions such as old/new and topic/comment. Japanese has many case particles and adverbial particles such as *ga* and *wa*. Topic Marker (TOP) *wa* is considered to carry old information and Focus Marker/Nominative (NOM) *ga* marks new information. A piece of evidence is found in (11Q), where *ga*, not *wa*, is compatible with WH *dare* "who."

(11) Q: Dare-ga/*wa   pati-ni   ki-mashi-ta-ka?
          who-NOM/*Top   party-LOC   come-POLITE-PAST-Q
          "Who came to the party?"

      A1: John-ga   ki-ta.   (neutral *ga*)
              NOM   come-PAST
          "John came."

      A2: John-wa   ki-ta.   (contrastive *wa*)
              TOP   come-PAST
          "As for John, he came."

The distinction between new and old information is morphologically cued in particles *ga* and *wa* in Japanese and it is expected that the old/new distinction does not need to be marked in nouns by local F0 boost. We will conduct a production and perception experiment to see how information new focus is marked in chapter 3.

Note here that (11A2) is also a possible answer to (11Q). Hara (2006) observes that contrastive *wa*, as in (11A2), implicates that the speaker does not know about other individuals but she only asserts a proposition on somebody who is contrasted. Tomioka (2009) observes that contrastive *wa* gets prominent.

(12) Q: Dare-ga   ukat-ta-no?

Who-GA   pass-PAST-Q

"Who passed?"

A: KEN-wa/Ken-WA   ukat-ta.

Ken-TOP   pass-PAST

"(At least) Ken passed."   (Tomioka, 2009: 119)

Tomioka (2009) claims that either subject *KEN* or particle *WA* gets prominent in (12A). We will come back to the difference in meaning in chapter 7.

Information structure notions and prosody have been much studied independently in Japanese literature, but the interrelation between them is understudied. Usually, information particles like *ga* and *wa* appear without lexical prominence, but there are cases where these particles are prominent. Prominence is imposed not only on content words but also on function morphemes in Japanese (cf. figures 1.2–1.3 in chapter 1).

Kuno (1973) distinguishes two types of *wa*: thematic *wa* (cf. (13a)) and contrastive *wa* (cf. (13b)).

(13) a. Thematic *wa*

John-wa   gakusei-desu.

TOP   student-copula

"John is a student."

b. Contrastive *wa*

John-ga   pai-wa   tabe-ta-ga   keki-wa

NOM   pie-TOP   eat-PAST-but   cake-TOP

tabe-nakat-ta.

eat-NEG-PAST

"John ate (the)[3] pie, but he did not eat (the) cake."

Kuno (1973) names the use of *wa* as in (13a) "thematic *wa*" and it marks "theme," in the sense of the Prague School. Thematic *wa* does not encode any sense of contrast and is sometimes called "non-contrastive *wa*" in Japanese literature (cf. Heycock, 1994). "Contrastive *wa*," on other hand, conveys contrast, as observed in (13b). Kuno (1973) points out that contrastive *wa* can iterate in a sentence, while thematic *wa* cannot. In (14), the first wa_1 is thematic *wa*, but the second and the third *wa* are contrastive.

(14) Watashi-<u>wa</u>$_1$ sake-<u>wa</u>$_2$ nomi-masu-ga   tabako-<u>wa</u>$_3$
       I-TOP   alcohol-TOP   drink-POLITE-but   tabacco-TOP
       sui-mase-n.
       smoke-POLITE-NEG
       "I drink, but (I) do not smoke."

Particle *ga* also has two readings: neutral and exhaustive, as shown in (15) (cf. Kuno, 1973).

(15) a. Neutral *ga*
           Ame-ga   hut-tei-masu.
           rain-GA   fall-PROG-POLITE
           "It is raining."
      b. Exhaustive *ga*
           John-ga   gakusei-desu.
           GA   student-copula(POLITE)
           "(Of all the people under discussion) John (and only John) is a
           student."

Whether *ga* has the exhaustive reading or not depends on the context. Kuroda (1965) argues that individual-level predicates are compatible with the exhaustive listing (cf. (16a)), but stage-level predicates either have the neutral or the exhaustive listing (cf. (16b)) with a *ga*-marked subject.

(16) a. John-ga   gakusei-desu.   (exhaustive)
           GA   student-copula
           "John (and only John) is a student."
      b. John-ga   ki-ta.   (neutral/exhaustive)
           GA   come-PAST
           "John came. /John (and only John) came."

It is worth noting again that, in Japanese, function morphemes like *wa* are assigned focal prominence. Kuno (1973) points out that contrastive *wa*, not thematic *wa*, is associated with prominent intonation (cf. (12A)). Generally speaking, in English and other Germanic languages, prominence is marked on content words, not on function words (cf. Baumann and Winter, 2018; Bishop et al., 2020). In Japanese, that is not the case. Note that *wa* is not the only particle to get prominent; *ga* in (17A), for example, is also associated with prominence when it has an exhaustive reading. Neutral *ga*, on the other hand, does not mark prosodic prominence.

(17) Q: Dare-ga   pati-ni   ki-ta-no?

       who-NOM   party-to   come-PAST-Q

      "Who came to the party?"

  A: John-ga/-GA   ki-ta-yo.

      NOM   come-PAST-Particle

      "John came (neutral reading) / Only John came (exhaustive reading)."

Oishi (1959) points out that case particles like *ni* "to," *de* "in/at" as well as adverbial particles like *made* "even," *sae* "even," and *mo* "also" can also receive prominence depending on the context. Meanings are different between particles with and without prominence in (18).

(18) a. Pati-ni/NI John-ga/GA ki-ta.   Mary-mo/MO   ki-ta.

     party-to   GA   come-PAST   too   come-PAST

     "To the party, John came. Mary came, too."

  b. A: Gakko-NO   toronkai?

       school-GEN   debate

      "A debate ABOUT the school?"

    B: Iya, gakko-DE   yat-ta-n   desu.

      no   school-AT   do-PAST-COMP   copula (Polite)

      "No, (we) did a debate AT the school."   (Oishi, 1959: 100)

Japanese marks function morphemes like particles prosodically prominent. However, having prominence on a particle does not mean that the particle is in contrast with other particles. The domain of prominence is larger than the particle itself. In Rooth's Alternative Semantics (see chapter 7 for details), (17A) evokes an alternative set {John came to the party, Mary came to the party, Sarah came to the party . . .}, and what the speaker of (17A) asserts is that it is John who came, and implicates that she does not know about other people. The implicature is ready to be canceled; (17Aa) can be followed by a sentence like *Mary-mo ki-ta* "Mary came, too."

In section 2.1, we introduce FoC proposed by K & S (2020). It is classed as a contrastive focus that induces pragmatic implicature. Focal prominence on particles in Japanese induces pragmatic implicature and we claim that Japanese has FoC.

In this section, we have reviewed and discussed the kinds of focal prominence that Japanese realizes prosodically. We suspect Japanese probably does not mark broad focus. Narrow focus, the anti-notion of broad focus, is a cover term for corrective focus, contrastive focus, FoC, and information new focus in the traditional focus literature. We will conduct production and perception experiments and see what kinds of focal prominence Japanese marks prosodically in chapters 3, 4, and 6. We believe that perception experiments are indispensable to see whether focal prominence has an effect on utterance comprehension by a listener.

NOTES

1. In this representation of tone patterns, we follow the custom of traditional Japanese linguistics in which each tone is linked to each mora.

2. The accent *HL is not recognized in a Final-Accented word without a following mora/syllable, so we put (-L) here to make it clear that word is accented.

3. Japanese lacks definite and indefinite articles. We tentatively insert "the" here.

Chapter 3

Focal Prominence on Lexical Word

This chapter investigates how the focal prominence of a lexical word is produced and perceived in Japanese. As reviewed in chapter 2, Pierrehumbert and Beckman (P & B) (1988) is a pioneering work on focal prominence (cf. figure 2.1 in chapter 2). Many focus studies followed their work and proposed that Japanese has the so-called "focus effects": focal F0-rise and post-focal compression. Most of them, however, do not define "focus" and lack quantitative confirmation (cf. Ishihara et al., 2018). We will investigate what type of focus Japanese marks by production experiments above the word level. We asked our participants to produce the materials as a Question-Answer pair so that the answers to each WH question contain a broad focus (cf. (1a)) and an information new focus (cf. (1b,c)), respectively.

(1)   a.   What is it? Type (gives the broad focus context)

         Q: Nani-ga   suki-desu-ka?

            what-NOM   like-Copula(POL)-Q

            "What (do you)[1] like?"

         A: Aoi   uni-desu.

            blue   sea urchin-Copula(POL)

            "(I like) blue urchin."

     b.   What Y was it? Type (gives the information new focus context on Modifier (cf. (2))

         Q: Donna   uni-ga   suki-desu-ka?

            what   sea urchin-NOM   like-Copula(POL)-Q

            "What sea urchin (do you) like?"

A: Aoi   uni-desu.

blue   sea urchin-Copula(POL)

"(I like) blue [+F] urchin."

c. X's what was it? Type (gives the information new focus context on Head)

Q: Aoi   nani-ga   suki-desu-ka?

blue   what-NOM   like-Copula(POL)-Q

"What blue (do you) like?"

A: Aoi   uni-desu.

blue   sea urchin-Copula(POL)

"(I like) blue urchin [+F]."

Another difference between our experiments and the traditional analyses is that we take the normalized F0-means of the six measurement points to factor out pitch range differences among collaborators and obtain more objective data than figure 2.1 provided by P&B (1988).

A third difference is that we have conducted a perception experiment of focal prominence. The previous studies on Japanese prosody are mostly production studies. We wonder whether focal prominence is effective in utterance perception. We believe we need to know how prominence is perceived by experiments.

The structure of this chapter is as follows; section 3.1 is on our production experiment and section 3.2 is on our perception experiment. Discussion follows in section 3.3.

3.1 PRODUCTION EXPERIMENT

3.1.1 Stimuli and Method

Our stimuli consisted of two words with similar segmental characteristics, following P&B (1988) (cf. (2)). They cover the four possible accentual sequences as in (3): AA, AU, UA, and UU, where U is an unaccented word, and A is an accented word.

(2)   The Structure of Stimuli

$[_{\text{Phrase}} [_{\text{Mod}} \text{XXXX}][_{\text{Head}} \text{YYYY}]]$-desu

(3)   AA: /a'oi 'uni 'desu/   "It is a blue sea urchin."[2]

blue sea urchin Copula(POL)

/a'oyama-no 'ani 'desu/ "It is my brother in Aoyama."

(place name) brother Copula(POL)

AU: /a'oi ume 'desu/ "It is a blue plum."

   plum

/a'oyama-no ane 'desu/ "It is my sister in Aoyama."

    sister

UA: /amai 'uni 'desu/ "It is a sweet sea urchin."

   sweet

/oomiya-no 'ani 'desu/ "It is my brother in Omiya."

(place name)

UU: /amai ume 'desu/ "It is a sweet plum."

/oomiya-no ane 'desu/ "It is my sister in Omiya."

We included a [+Focus] condition regarding which word in the sequence was focused: the first word, the second word, or neither. This yields 12 stimuli conditions: UU, U[+F]U, UU[+F], UA, U[+F]A, UA[+F], AU, A[+F] U, AU[+F], AA, A[+F]A, and AA[+F]. The context for focus was given as answers to a wh-question: broad focus and information-new focus on W1 and W2 (cf. (1)).

Five Tokyo-Japanese speakers (M3, F2) participated in the experiment. We recruited eight participants, but the data of the three speakers were abandoned because their lexical accents were not consistent. Recordings were conducted in a sound-attenuated booth at a Japanese university and were saved directly to a laptop computer at 44.1kHz with 16 bits per sample. During the recordings, stimuli were presented to participants through PowerPoint slides in a pseudo-randomized order. Participants, watching a computer display, first silently read a question that gives the context. After that, they read the stimulus sentences aloud as answers to the contextual questions. Twenty-four stimuli and 24 distractors (see appendix A for the materials) were produced per participant. The experiment including the exercise session took about 30 minutes to complete.

For the analysis, we follow Ishihara (2016) and take the normalized F0-means of the six measurement points: the 1st F0-minimum, the F0-maximum, and the 2nd F0-minimum of Word 1 (L1-1, H1, L1-2) and the same measures for Word 2 (L2-1, H2, L2-2). The data from all five participants were normalized to factor out pitch range differences among speakers, using the formula in (4), where x is an actual value in each participant's data, R1 is the F0 peak, and R2 is the F0 valley of the whole utterance.

(4) Normalization formula: $y = (x - R2) / (R1 - R2)$

3.1.2 Results and Discussion

Figure 3.1 shows the normalized F0 of two-word sequences of broad focus condition (cf. (1a)), where L1-1 is Initial Low %L and H1 is accentual or phrasal H.

The H1s in AA and AU are higher than the H1s in UU and UA, and we can see the accentual boost (cf. Shinya, 2009) here. The H2s in AA and AU are lower than the H2s in UU and UA, which shows that A-words lower the F0 pattern following them, known as "catathesis (a.k.a. downstep)." In two-word sequences in the form of structure (3), the maximum F0 is located on H1 of the first word, and H2 of the second word is always lower than H1, irrelevant to the accentual patterns of words. Figure 3.1 shows that the F0 lowers over the course of an utterance and we observe a downtrend here.

Figures 3.2 through 3.5 show the measurements of mean normalized F0 in each of the four two-word accentual conditions (AA, AU, UA, and UU, respectively), with the [+F] condition added.

The literature reports that F0 is boosted when focused (cf. Ishihara 2016, among others), but the F0 boost in our data set is relatively small; only the H2 in UA[+F] and AU[+F] are significantly higher than the H2 in UA and AU (w=7, p=0.026, for both conditions, on Mann-Whitney test) out of 8 focus positions of two-word sequences (cf. figures 3.3–3.4). In UA[+F] condition, the H2 with [+F] in the second word is even higher than the H1 in the first word. This is the only position where focused H2 is higher than H1

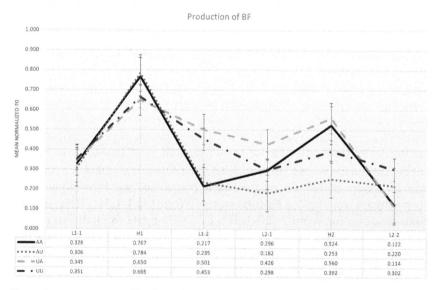

	L1-1	H1	L1-2	L2-1	H2	L2-2
AA	0.328	0.767	0.217	0.296	0.524	0.122
AU	0.306	0.784	0.235	0.182	0.253	0.220
UA	0.345	0.650	0.501	0.426	0.560	0.114
UU	0.351	0.665	0.453	0.298	0.392	0.302

Figure 3.1   Mean Normalized F0 of Two-Word Sequence in Broad Focus Condition.

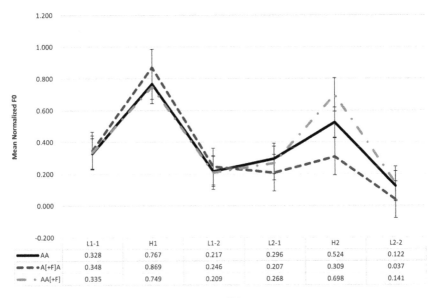

	L1-1	H1	L1-2	L2-1	H2	L2-2
AA	0.328	0.767	0.217	0.296	0.524	0.122
A[+F]A	0.348	0.869	0.246	0.207	0.309	0.037
AA[+F]	0.335	0.749	0.209	0.268	0.698	0.141

Figure 3.2   Mean Normalized F0 in AA Condition.

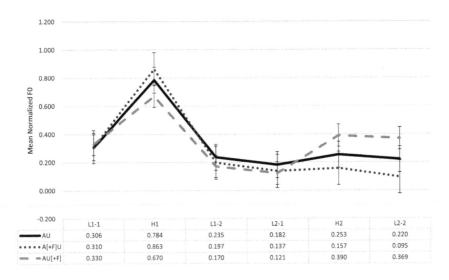

	L1-1	H1	L1-2	L2-1	H2	L2-2
AU	0.306	0.784	0.235	0.182	0.253	0.220
A[+F]U	0.310	0.863	0.197	0.137	0.157	0.095
AU[+F]	0.330	0.670	0.170	0.121	0.390	0.369

Figure 3.3   Mean normalized F0 in AU condition.

(cf. figure 3.4). In the other focus environments, that is, A[+F]A, AA[+F], A[+F]U, U[+F]A, U[+F]U, UU[+F], U[+F]A, the H1 in the first word is higher in F0 than the H2 in the second word, irrelevant of the accentedness of the first and the second words.

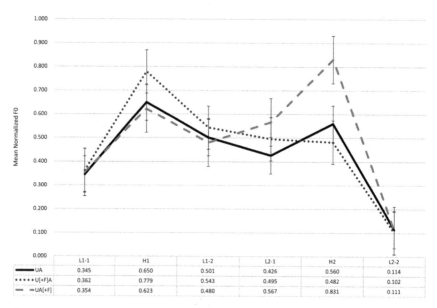

Figure 3.4   Mean normalized F0 in UA condition.

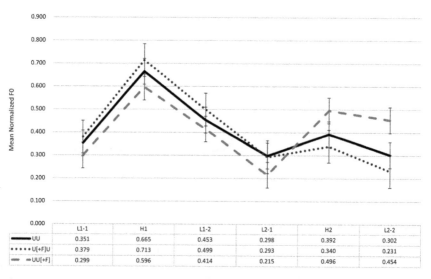

Figure 3.5   Mean normalized F0 in UU condition.

Post-focal compression is another focus-related F0 change reported in the Japanese focus literature (cf. Ishihara 2016, among others), but in our data set, the H2 in A[+F] U is the only position (cf. figure 3.3) where H2 is significantly compressed than the H2 in AU (w=41, p=0.038 on Mann Whitney test).

Through our production experiments and analyses, we have made clear the phonetic properties of focal prominences in Japanese and confirmed P&B's informal F0 contours in figure 2.1 in chapter 2 statistically. Overall, though F0-boost is observed, it is significantly different from non-focused H only in AU[+F] and UA[+F]. In other focused environments, A[+F]A, AA[+F], A[+F]U, U[+F]A, U[+F]U, and UU[+F], the focus is boosted but the differences between focused H and non-focused H are not significant. Post-focal compression is significantly observed in the A[+F]U position only. We conclude that the focus effects of F0-boost and post-focal compression are not as strong as claimed in the traditional studies.

As the prosodic features characteristic of focus failed to show up constantly in the conditions considered, we suspect that the F0 cues and the focal prominence are not correlated so straightforwardly in Japanese. If the effects on focus in terms of prosody are categorical, we expect that listeners can identify focus easily. We will examine whether listeners can perceive focus or not through a perception experiment in the next section.

3.2 PERCEPTION EXPERIMENT

3.2.1 Methods and Predictions

We tested the perception of focus in the same 12 stimuli contexts as in our production experiment: UU, U[+F]U, UU[+F], UA, U[+F]A, UA[+F], AU, A[+F]U, AU[+F], AA, A[+F]A, and AA[+F]. We also examined the perception of broad focus by providing stimuli produced under broad focus and providing a "broad focus" response option. We expect that listeners will choose the "broad focus" response option both when an utterance was produced with a broad focus and when the acoustic cues to focus are unclear.

Twenty-three Japanese students (M11, F12, mean age 19.65, SD 4.0) were recruited as participants at a Japanese university. They reported no auditory difficulties. Before the experiment, they received instruction on the relation between WH questions and focus, including broad focus (BF) and information-new focus (IF), based on (1). In the experiment, they only heard answers without texts and were asked to mark which position in the target sequence was focused on the answer sheet. They were instructed to mark broad focus as "The Whole Utterance." Each participant listened to 96 tokens of utterances: 48 with the 12 accentual and focal patterns (4 per pattern), 24 with a broad focus, which was collected in the production experiment, and 24 distractors (tokens with numbers) (see appendix B for materials). The experiments were conducted in a quiet room at a Japanese university, and the audio stimuli were presented to listeners through a room

Table 3.1   Focus Identification Rates (Percentage Values)

Focus type	Max (%)	Min (%)	Mean (%)	SD
IF on A-words	91.4	68.5	78.8	16.1
IF on U-words	76.1	57.6	65.8	4.6
BF	42.4	15.3	38.6	5.4

(where A-words mean "accented" words, and U-words mean "unaccented" words)

speaker. Before the actual test, we presented three practice trials to participants to familiarize them with the procedure. Each session lasted about 30 minutes.

We predicted that since the so-called "focus effects" found in our production experiment are not strong in Japanese, participants would have difficulties perceiving focus. Also, as the experiment requires listeners to choose among the three choices, IF on the first word, IF on the second word, and BF, we predicted that the task would not be easy.

3.2.2 Results

Table 3.1 exhibits the identification rates of BF and IF; columns correspond to focus positions within an utterance, and rows represent the focus identification rates by percentage values.

The overall identification rate was 78.8% for IF on A-words, 65.8% for IF on U-words, and 38.6% for BF. To test whether the numerical difference among identification rates was statistically supported, we conducted a one-way ANOVA between focus type and identification rates, and the result was $F_{(2, 275)} = 8.618$, $p = 0.0002$. A post hoc multiple comparison analysis showed a significant difference between BF and IF on A-words ($p < 0.001$) as well as between BF and IF on U-words ($p = 0.004$), but not between IF on A-words and IF on U-words ($p=0.29$).

Table 3.2   Confusion Matrix of BF and IF (Percentage Values)

	Target				Target		
Perceived	A[+F]A	AA	AA[+F]	Perceived	A[+F]U	AU	AU[+F]
A[+F]A	68.5	38	9.8	A[+F]U	77.2	51.1	16.3
AA	21.7	15.3	12	AU	22.8	31.5	18.5
AA[+F]	9.8	46.7	78.2	AU[+F]	0	17.4	65.2

	Target				Target		
Perceived	U[+F]A	UA	UA[+F]	Perceived	U[+F]U	UU	UU[+F]
U[+F]A	57.6	23.9	4.3	U[+F]U	64.1	37	10.9
UA	23.9	42.4	4.3	UU	26.1	21.7	13
UA[+F]	18.5	33.7	91.4	UU[+F]	9.8	41.3	76.1

Though the overall identification rates are not significantly different between A[+F] and U[+F], a closer look shows a different picture. Table 3.2 is a confusion matrix of IF and BF perception, and we can see the identification rates of IF vary from 57.6% to 91.4%, depending on focus positions.

Another finding is that the identification rates of BF are surprisingly low both in tables 3.1 and 3.2.

3.3 DISCUSSION

Japanese uses pitch to mark lexical accent (cf. figure 1.1 in chapter 1) and intonation (cf. figure 1.2 in chapter 1). Accented (A-) words show a higher F0 peak than Unaccented (U-) words, which is called "Accentual boost" in the literature (cf. figure 3.1, Shinya 2009)). A-words are considered different from U-words in that they lower the F0 pattern following them, known as "catathesis" (a.k.a. downstep) (cf. figure 1.1, figure 3.1) in Japanese literature. Generally, it is assumed in the prosody study that the higher the F0, the more prominence it yields (cf. figure 3.6(A)).

On this assumption, it is predicted that A-words would be perceived with greater prominence than U-words and that a downstepped word after A-word is less prominent than a non-downstepped word after a U-word. Our perception experiment, however, shows that neither is borne out; overall, there is no significant difference between A-words and U-words in focus perception rates (cf. table 3.1). Focus after A-words is not perceived less than focus after U-words, either (cf. table 3.2); the identification rates are 91.4% in UA[+F], 78.2% in AA[+F], 76.1% in UU[+F], 65.2% in AU[+F], and the difference in perception rates is not significant between UA[+F] and AA[+F] on T-test ($t(44)=1.778$, $p=0.08$), nor between UU[+F] and AU[+F] ($t(44)=0.149$, $p=0.882$). What is interesting about our experiment results is that even when the F0 of the second word is boosted by focus, its F0 peak is lower than the F0 peak of the first word that is not focused, except in the UA[+F] condition (cf. figures 3.2–3.5), but the second F0 peak is perceived more prominent than

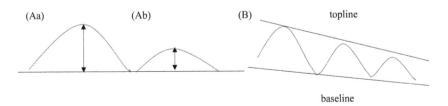

Figure 3.6 (A) Excursion Size in F0, (B) Topline and Baseline of an Utterance Contour.

the first F0 peak (cf. table 3.2). We find that the relation between F0 peak and prominence is not straightforward in Japanese.

The literature (cf. Pierrehumbert 1979; Gussenhoven et al. 1997, among many others) says that prominence perception is affected by the excursion size of the F0 peak. In figure 3.6, for example, contour (Aa) invokes more prominence than contour (Ab). It is also known that the location of the F0 peak affects perceived prominence. Pierrehumbert (1979) showed, based on perception experiments on the *mamamama* nonce sequence, that when two F0 peaks sound equivalent, the second F0 must be lower. She proposed a perceptual process that normalizes declination; listeners know that an F0 peak that appears later in an utterance is lower than an earlier peak due to declination, and compensates for it in perception. Excursion and declination are incorporated into a prominence theory by assuming F0 peaks and valleys in an utterance; the former forms the "topline" and the latter forms the "baseline," as illustrated in figure 3.6 (B). As the utterance goes on, both the topline and the baseline lower so that F0 peaks are getting lower, accordingly.

In the normalization theory, it is assumed that F0 peaks appearing later sound more prominent than the earlier F0 peaks, even though it is lower in pitch. It looks to explain our experiment results, but Japanese utterance contours are more complex than the ones in figure 3.6(B). Japanese uses pitch on various levels of prosodic phrases; in the traditional analyses (cf. chapter 1), lexical bi-tonal H*+L accent is specified on the lexical level, and a phrase boundary is marked by the initial low %L in Tokyo Japanese, φ-phrase is the domain of catathesis, and the Intonational Phrase (IP) level and the Utterance level realize downtrend (a.k.a. declination), that is, the downward pitch movement over the course of the utterance. Japanese also aligns boundary tones H%, HL%, and LH% at the phrase final position (cf. figure 1.5 in chapter 1, Venditti et al., 2008). As will be discussed in chapter 7, these boundary tones invoke pragmatic implicature. It is expected that pitch movements on different levels interact somehow and impact the perception of prominence.

Shinya (2009) investigated the effects of lexical accent on the perception of prominence in Japanese by perception experiments; he used the synthesized pitches in two-word sequences AA, AU, UA, and UU, and by gradually changing F0s, he investigated the interaction between the two words when they sound equally prominent. His main finding is that two distinct perceptual processes are at play in Japanese: "accentual boost normalization" and "downstep normalization." Shinya (2009) demonstrates that when an A-word and a U-word are perceived as having the same prominence, the A-word has a higher F0 peak value than the U-word. In the Japanese prosody, the accentual boost of A-words needs to be normalized so that the extra F0 boost assigned by a lexical accent does not count as part of the F0 peak's excursion. The other normalization is downstep normalization; it compensates for the production effect of the downstep. For an F0 peak to be perceived as having equivalent prominence to a preceding F0 peak, the second peak is always lower in F0 when the first word is accented than when it is unaccented. Shinya's finding is important to point out that F0 peaks are normalized in Japanese when perceived. In other words, the mean F0 peak cannot be a proper prominence perception cue as is.

Returning to our experiment data, we see that Shinya's two types of normalization are compatible with our results. Table 3.1 shows no significant difference in the identification rates between focused A-words and focused U-words, and this is explained by Shinya's "accentual boost normalization." His "downstep normalization" explains the fact that the first F0 peak is higher than the second one, but still, there are cases where the second focused F0 peak is perceived as prominent, as observed in figures 3.2–3.5, and table 3.2. Shinya's work is important to show that lexical accent affects pitch movement on a phrase level higher than the lexical level and is normalized when perceived. Pierrehumbert's work is also important to propose a perceptual process that normalizes declination on a higher phrase level than the lexical level.

This chapter aims to investigate focal prominence on the phrase level. Our research questions were whether or not Japanese has a "broad" focus and 'information new' focus, and how focal prominence, if any, is produced and perceived in Japanese. Based on production experiments, we conclude that Japanese marks information-new focus. Its production cue is a local F0 boost but we have found that it is not as strong as claimed in the traditional literature. As for broad focus, we do not consider Japanese aligns a broad focus. Our experiment materials in this chapter are composed of lexical pitch accents and phrasal tones, and we cannot find a systematic assignment of broad focus (cf. figure 3.1). The perception rates to recognize prosody as a broad focus are low (cf. table 3.1). This may be because listeners do not recognize broad focus, and this leads us to conclude that Japanese does not mark a broad focus in this chapter.

Japanese prosody is formed compositionally by the lexical pitch accents, phrasal tones, and boundary tones (Féry, 2017: 248). Our experiments in this chapter show that focal prominence is marked by a local F0 boost, but the lexical F0 boost is normalized upon perception. If we are on the right track to claim that lexical accent does not affect the perception of prominence much, the next issue to consider is how phrasal tones affect the Japanese prosody and the perception of prominence. In chapter 4, we will work on focal prominence not induced by lexical accents.

NOTES

1. Japanese can delete arguments as far as they are recoverable from the context. We put *I/you* for the subject in the gloss tentatively.

2. For the marking of accent, we adopt an IPA stress symbol at the beginning of the accented mora.

Chapter 4

Focal Prominence without Lexical Accent

In chapter 3, we discussed broad focus and information-new focus; based on production and perception experiments on lexically accented content words, we conclude that Japanese does not mark broad focus. Information-new focus is marked by local F0 boost, though the boost is not so strong. This chapter discusses corrective focus in Japanese. Corrective focus is a kind of narrow focus and is cued by a local focal prominence on the word to be corrected. The materials in our experiment in this chapter are numbers. They are lexically accented when uttered independently, but Tokyo Japanese has a special prosodic pattern for telephone numbers (Lee et al., 2019). This chapter aims to see whether not only local F0 boost but also phrasal tones are effective to mark focal prominence in Japanese.

We will review a cross-linguistic variation of corrective focus first, and then consider Japanese. We conducted a perception experiment and found that Japanese corrective focus is not always easy to hear even to L1 listeners, due to the mismatch between the phrasal tone and a local F0 boost.

4.1 CROSS-LINGUISTIC VARIATION OF CORRECTIVE FOCUS

Cross-linguistically, the focus is marked segmentally, tonally, syntactically, or is not marked at all (Kratzer and Selkirk (K&S) 2020). It is a natural recourse that focus is perceived not in a single strategy. Lee et al. (2015) work on the production and perception of corrective focus in seven languages and dialects and show that acoustic cues of a language account for the perception of corrective focus cross-linguistically, using a paradigm based on 10-digit number strings in different languages. They ask subjects to read digit strings in isolation as a background broad focus condition, and as an answer in a Q&A dialogue like (1) for corrective focus.

(1)   Q: Is Mary's number 215-418-5623?
        A: No, the number is 215-417-5623.

     They recruited a total of 32 subjects (5 American English (AE), 5 Mandarin Chinese (MC), 5 Seoul Korean (SK), 5 South Kyungsan Korean (SKK), 6 Suzhou Wu (Wu), 3 Tokyo Japanese (TJ), and 3 Standard French (SF) speakers), and measured the mean F0, duration and intensity of focused digits. Table 4.1 shows the median values of focused digits.

     Lee et al. (2015) further conducted a perception experiment of corrective focus in Seoul Korean, South Kyungsan Korean, Mandarin Chinese, and American English. The accuracy rate to identify the corrective focus is given in the last row of table 4.1. They found differences among languages in how prosodic focus is acoustically marked in production and is recognized in perception; English and Chinese, for example, have strong acoustic cues and the focus identification rates are high. Seoul Korean and South Kyungsan Korean foci are neither clearly marked in production nor accurately recognized in perception. Lee et al. (2015) claimed that for languages with a stronger acoustic marking of corrective focus, L1 listeners were able to identify the focused material with almost perfect accuracy (97.3% for American English and 94.9% for Mandarin Chinese). On the other hand, in languages with a weaker marking of corrective focus, L1 listeners were able to identify the corrected number above chance but at very low accuracy rates (44.6% for Seoul Korean and 55.6% for South Kyungsan Korean).

     Lee et al. (2015) classify Tokyo Japanese as the language group with weak acoustic cues, based on their production experiment (cf. table 4.1). They, however, did not conduct a perception experiment and the focus identification rate in Japanese was unknown.[1] Below we will conduct a perception experiment on Japanese corrective focus and see whether their claim is borne out.

Table 4.1   Median (z-score) Values of Focused Digits and Perception Rates (cited from Lee et al., 2015 with revision)

	SK	SKK	TJ	Wu	SF	MC	AE
Duration	0.13	0.64	0.10	0.48	1.73	1.19	0.95
Intensity	0.24	−.26	−.24	0.53	0.97	0.36	1.28
Pitch	0.62	1.00	0.60	0.61	1.17	3.13	2.96
Perception rate	44.6%	55.6%	–	–	–	94.9%	97.3%

weak ⟵————————————————⟶ strong
prosodic focus marking

4.2 PERCEPTION EXPERIMENT ON JAPANESE CORRECTIVE FOCUS

4.2.1 Stimuli and Method

We followed the methodology used by Lee et al. (2015). One male speaker of Tokyo Japanese recorded a series of 10 question-answer pairs which contained a 10-digit number of the form XXX-XXX-XXXX (cf. (2)) in a sound-attenuated room. In each of the responses, one of the 10 numbers was produced with a corrective focus.

(2)   Japanese material

  Q: Yamada-san-no   denwa-bango-wa 215-413-5623 desu-ka?

    Yamada-Mr/Ms-GEN phone-number-TOP   COPULA-Q

    "Is Yamada's number 215-413-5623?"

  A: Iie. 215-413-6623 desu.

    No.   COPULA

    "No. (It)[2] is 215-413-6623."

Tokyo Japanese has a special prosodic pattern for telephone numbers (Lee et al., 2019). In a phone number string, all digits become two morae; *ni* "two" and *shi* "four," for example, become ˈ*nii* and ˈ*shii*, as given in (3).

(3)   ˈ*zero* (zero), ˈ*ichi* (one), ˈ*nii* (two), ˈ*san* (three), ˈ*yon*/ ˈ*shii* (four), ˈ*goo* (five), ˈ*roku* (six), ˈ*nana*/*shichi* (seven), ˈ*hachi* (eight), ˈ*kyuu* (nine)

A 10-digit phone number string comes in three prosodic phrases in Japanese: $_{\varphi 1}[N_1\ N_2\ N_3]\ _{\varphi 2}[N_4\ N_5\ N_6]\ _{\varphi 3}[_{\varphi minor\ 1}[N_7\ N_8\ ]_{\varphi minor\ 2}[N_9\ N_{10}]]$ (where N is number). The first and the second φ-phrase[3] have one F0 peak, and the third φ-phrase has two. Observe figure 4.1, which shows the pitch movement of the 10-digit number 222-222-2222 without a focal prominence, recorded by one of the authors.

We recruited three groups of listeners, one group of L1 listeners and two groups of L2 listeners, to see a language bias. 22 (M2, F20) Japanese L1 listeners between the ages of 18 and 24 (mean age 20.45, SD=0.87) were recruited at two Japanese universities, 17 Chinese students (M5, F12, mean age 25.59, SD=2.67) at a Japanese university, and 9 Dutch students (M4, F5, mean age 24.04, SD=5.04) at a Belgian university. All of the L2 listeners

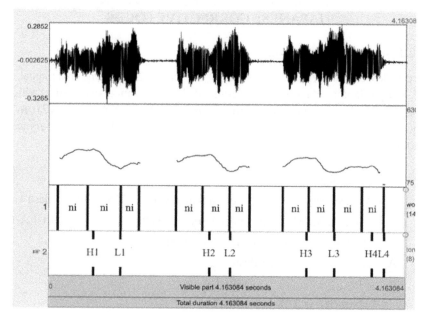

Figure 4.1   Pitch Movement of Ten-Digit Number 222-222-2222 in Japanese.

were advanced learners of Japanese. Our participants reported no hearing difficulties.

In the experiment, participants listened to each recording twice (see appendix C for the experiment materials) and marked the number that they perceive as prominent on a transcript. They listened to 30 randomized phone-number strings, that is, each corrective digit was repeated three times, in a quiet room through a room speaker. The experiment, including the exercise session, took about 30 minutes to complete.

4.2.2 Results

The identification rate of Japanese corrective focus by 22 Japanese L1 listeners was 86.2%. Table 4.2 is the confusion matrix in percent accuracy.

Though Lee et al. (2015) consider that Korean and Japanese are the languages with a weaker marking of corrective focus (cf. table 4.1), the focus identification rate of Tokyo Japanese is much higher in our experiment than those of Korean: Tokyo Japanese (86.2%) > South Kyungsan Korean (55.6%) > Seoul Korean (44.6%).

Table 4.2 Confusion Matrix of Japanese Corrective Focus Perception in Percentage Values by L1 Japanese Listeners

Perceived Target	1	2	3	4	5	6	7	8	9	10
1	100									
2	9	88	3							
3	3	2	91		2	2				
4				97	3					
5		2			98					
6			2	4		92	2			
7							65	33	2	
8			2		2	2	2	91	1	
9									82	18
10	3		2		2	3	3	3	27	58

Table 4.3 Confusion Matrix of Japanese Corrective Focus Perception in Percentage Values by Chinese Learners (Above) and by Dutch Learners (Below)

Perceived Target	1	2	3	4	5	6	7	8	9	10
1	100									
2		100								
3	2	4	92	2						
4	2			92	6					
5		2		2	96					
6				2		98				
7							80	18	2	
8			2				2	94	2	
9									69	31
10	2						2	2	4	90

Perceived Target	1	2	3	4	5	6	7	8	9	10
1	100									
2		96	4							
3	4	4	81	11						
4				100						
5				4	93	4				
6						100				
7							44	56		
8							7	85	7	
9							7		63	30
10		4							4	93

Table 4.4 Accuracy of Perception of Corrective Focus in Japanese and English by Japanese Listeners

	Target Number Position (% accuracy)									
	1st φPhrase			2nd φPhrase			3rd φPhrase			
	#1	*#2*	*#3*	*#4*	*#5*	*#6*	*#7*	*#8*	*#9*	*#10*
L1 (Japanese): 86.2%	100	88	91	97	98	92	65	91	82	58
L2 (English): 98.6%	100	100	100	100	100	98	100	98	96	94

The results of corrective focus perception experiments by L2 learners of Japanese are more or less similar to the one by Japanese L1 listeners; the focus identification rate is 90.5 % for Chinese learners and 85.5% for Dutch learners. Table 4.3 shows the confusion matrix of Chinese learners (above) and Dutch learners (below).

Comparing table 4.2 with table 4.3, we can see that L1 listeners and L2 learners have a similar perceptual strategy; generally, they perceive corrective foci well, but the third φ-phrase is difficult to hear. Especially Positions #7 and #9 are easily mistaken for Positions #8 and #10 by all three groups.

4.3 PERCEPTION EXPERIMENT ON ENGLISH CORRECTIVE FOCUS

As for English experiments, Lee et al. (2015) report that the identification rate of corrective focus of American English is 97.3% (cf. table 4.1). We replicated their experiment on Japanese learners of English. One male speaker of General American recorded a series of 10 question-answer pairs which contained a 10-digit number of the form XXX-XXX-XXXX (see appendix C for the experiment materials) in a sound-attenuated room. We recruited 18 intermediate and advanced learners of English (M13, F5, mean age 20.5, SD=1) at a Japanese university. They had no hearing difficulties and followed the same procedures as the perception experiment on Japanese.

The focus identification rate was 98.6%. We can say that English L1 listeners and L2 learners have a similar perceptual strategy. What is surprising to us is that for Japanese listeners, Japanese corrective focus is harder to perceive than English ones. Compare again the perception rates in table 4.4.

We naively believe that L1 is easier to hear, but our experiments show that this is not always the case; the perception rate of Japanese corrective focus is 86.2% and that of English is 98.6% by Japanese listeners. This means that focus perception is not directly affected by language proficiency. In the next section, we will try to answer why L1 is not easy to hear.

4.4 WHY IS L1 NOT EASY TO HEAR?

4.4.1 Acoustic Measures

We examined the acoustic measures of pitch, duration, and intensity of our experiment materials. We included maximum F0 (Max F0), minimum F0, mean F0, range F0, and intensity and investigated how well the highest value of these measures in an utterance coincided with the number carrying corrective focus. Table 4.5 shows the z-scores of duration, intensity, and Max F0 of our experiment materials at the focused positions. Our materials show smaller differences between American English and Tokyo Japanese than those in Lee et al. (2015) (cf. table 4.1).

Figure 4.2 gives the normalized pitch and duration per focus position in our experiment materials in English and Japanese, and we can see clear differences between the two languages.[4] As mentioned above in section 4.2.1, a 10-digit phone number string comes in three φ-phrases in Japanese: $_{\varphi 1}[N_1\ N_2\ N_3]\ _{\varphi 2}[N_4\ N_5\ N_6]\ _{\varphi 3}[_{\varphi minor 1}\ [N_7\ N_8\ ]_{\varphi minor 2}[N_9\ N_{10}]]$. English 10-digit phone numbers also come in three phrases. As figure 4.2 (right) shows, the phrase-final focus has a longer duration at #3, #6, #8, and #10 in English. Assuming that φ_3 consists of two minor φ-phrases,[5] we can see that English realizes a longer duration at the phrase-final position but this is not the case for Japanese. Japanese does not look like using duration to mark corrective focus.

Another difference is observed in the use of Max F0. Figure 4.2 (left) shows that the Max F0 gradually declines from #1 to #10 in English. Japanese shows a completely different picture; #3 shows a sharp decline and the Max F0s at #7–#10 in φ_3 are very low. This may be "final-lowering" (cf. Maekawa 2011), typically observed at the utterance-final in Japanese.

Figure 4.3 describes other differences between English and Japanese; Max F0 corresponds with the position of the foci in English but not in Japanese. Max F0 does not necessarily match the focused segment; only three positions #2, #5, and #6 out of the 10 positions correspond with the focus position, and in the rest positions, Max F0 is not realized in the focus position.[6]

English is a stress language and focal prominence is marked by acoustic features of Max F0, duration, and intensity. Japanese is a pitch language, and its prosody is compositionally formed by lexical accents, phrasal tones,

Table 4.5 Median (z-score) Values of Focused Digits in English and Japanese

	General American English	*Tokyo Japanese*
duration	0.93	0.75
intensity	0.55	1.05
maximum F0	1.73	1.05

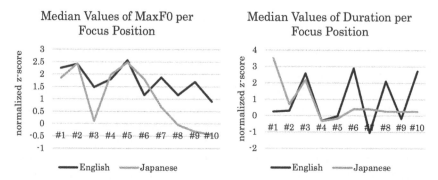

Figure 4.2   Median Values of Maximum F0 and Duration Per Focus Position in English and Japanese.

and boundary tones (cf. Venditti et al., 2008; Féry, 2017, among others). It is expected that Japanese focal prominence and acoustic features like F0 and duration are not straightforwardly related. We will consider why the mismatch occurs.

4.4.2 F0 Peak and Phrasal Tone in Japanese

The 10-digit telephone numbers, whether a number is accented (e.g. *'roku* "six") or unaccented (e.g. *san* "three") on the word level, form the same prosody. Recall figure 4.1 (repeated below as figure 4.4). Even if we replace

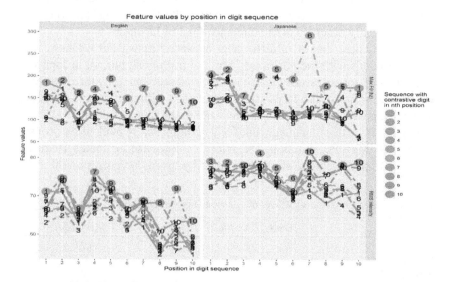

Figure 4.3   Maximal F0 and Intensity by Position in Digit Sequence (Left: English; Right: Japanese).

the number *nii* "two" with another number, the prosodic contour is the same as the one in figure 4.4. We can say that the pitch movement in figure 4.4 is formed not by lexical accents.

Let us decompose the pitch movement of the 10-digit number. We assume the 10-digit number has a phonological structure as in (4).

(4) $_{\varphi1}[N_1 \quad N_2 \quad N_3] \quad _{\varphi2}[N_4 \quad N_5 \quad N_6] \quad _{\varphi3}[_{\varphi minor\,1} \quad [N_7 \quad N_8]_{\varphi minor\,2} \quad [N_9 N_{10}]]$
%LH   HL   H%   %LH   HL   H%   %LH   HL%   H   HL%

(where %L and H% are boundary-initial and boundary-final tones)

(4) consists of three φ-phrases; the first and the second φ-phrase have one F0 peak, and the third one has two F0 peaks. In Tokyo Japanese, each φ-phrase begins with an initial low, that is, %L. It is followed by a phrasal H, which is aligned to N_1, N_4, and N_7. N_2 and N_5 are given a tonal peak H followed by L, irrespective of the lexical accent of the number itself.[7] In φ_1 and φ_2, phrase-final H% is assigned to N_3 and N_6, respectively. We assume φ_3 is composed of two minor phrases and each has its F0 peak, marked by HL. But unlike φ_1 and φ_2, no phrase-final H is aligned between the two minor phrases. As the tone movement in Tokyo Japanese is downward and F0 peaks lower as the utterance goes on, the HLs in φ_3 are lower than the ones in φ_1 and φ_2, due to declination (cf. figure 4.2 (left), figure 4.4 (left)).

We know that declination makes the F0 peaks lower which appear later in the utterance. Another declination effect is narrowing the range between a peak and a valley as the utterance goes on. In figure 4.4, for example, the pitch ranges between the peak and valley is broader in φ_1 and φ_2 than that in φ_3. We suspect that the lower F0 peak (cf. figure 4.2) and the narrower pitch range (cf. figure 4.4) are not strong focus cues, and the corrective foci in φ_3 are therefore difficult to perceive.

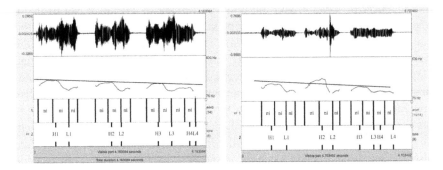

Figure 4.4 Pitch Movements of 10-Digit Numbers 222-222-2222, without Corrective Focus (left) and with Corrective Focus on #4 (right).

Japanese literature found the mismatch between a focus position and MaxF0 back in the 1980s. If we observe figure 4.4 closely, we recognize that the Max F0 marked as H1, H2, and H3 are located not on #1, #4, and #7, but are located on #2, #5, and #8. This is what Sugito (1980) calls *oso-sagari* "late fall." She found that pitch begins to lower a little bit later than the location of focus in Japanese, and she called this phenomenon *oso-sagari*. Despite the mismatch, the identification rates of #1 and #4 as corrective focus are nearly perfect (cf. tables 4.2–4.3). How can we perceive this mismatch?

4.4.3 Is There Any Perception Bias?

We have investigated possible prominence cues in English and Japanese. Acoustic cues of F0, duration, and intensity are effective to mark corrective focus as prosodically prominent in English. In Japanese, on the other hand, F0 peak and prosodic prominence are not straightforwardly related, since there are many factors to change F0 movements in an utterance. Catathesis and declination lower F0s as an utterance goes on (cf. McCawley, 1968; Poser, 1984; Ishihara, 2003, among others), and the F0 peak and the focus location do not always match because the pitch begins to lower a bit later than the location of a focus (cf. Sugito, 1980).

Pierrehumbert (1979) showed that the location of the F0 peak affects the perception of prominence. When two F0 peaks sound equivalent, the second F0 must be lower due to declination. Also, Shinya (2009) convincingly argued that accentual boost and catathesis are normalized when we perceive Japanese focus. In a word, F0 is normalized by the lexical accent and the location within an utterance when it is perceived. Mean F0 is not identified as is in Japanese, and it is normalized more or less. How do we then perceive F0s in Japanese?

There might be a Japanese-specific perception bias and Shport (2015) showed by perception experiments that Japanese listeners are sensitive to F0 fall. She investigated whether monolingual adult English listeners are sensitive to F0 peak location and the pitch movement of F0 fall novel to them. She conducted two types of perception experiments on resynthesized F0 patterns on non-words: discrimination task and categorization task. In the categorization experiment, listeners' sensitivity to the F0 peak location and F0 fall was examined in a prominence judgment task. She recruited 20 monolingual English listeners and 20 Japanese L1 listeners. Since both English and Japanese use post-lexical pitch movements like F0 fall, Shport (2015) predicted that both Japanese and English listeners are sensitive to the F0 fall in all oppositions, with a relatively small difference between the two groups. She also predicted that English listeners use F0 peak location over F0 fall as a cue to prominence, whereas Japanese listeners were predicted to use both cues, based on the previous works in the literature (cf. Gandour 1983; Kitahara

2001, among others). A relatively large difference between groups was expected, reflecting native language bias in pitch movement. Materials used in the two-alternative forced-choice categorization task are given in figure 4.5.

Shport (2015) used two acoustic parameters: F0 peak location and F0 fall after the peak. F0 peak location was varied in eight steps: 10%, 50%, and 90% in the first vowel [e], 50% in the second consonant [n], 10%, 50%, and 90% in the second vowel [e], and 50% in the third consonant [m]. At each location, the F0 fall either fell or did not fall after the peak in the discrimination task, and the F0 fall was varied in three steps, a large fall of 100 Hz, a moderate fall of 50 Hz, or no fall. The orthogonal variation of F0 peak location and F0-fall yielded 16 different F0 contours in the discrimination task and 24 F0 contours in the categorization task.

The stimuli given in figure 4.5 were repeated 10 times each, yielding 160 and 240 trials in each task. In the categorization task, listeners are required to determine whether stimuli in a pair were of the same category concerning the F0 fall. The total number of listeners' responses was 9,600. Figure 4.6 illustrates the results of the categorization task: left for the Japanese group and right for the English group. The main effect of the group and its interaction with F0 fall and peak location was significant (p<.036). F0 fall (F(2,4776)=150.81, p<.001), peak location (F(7, 4776)=88.38, p<.001), and interaction between these two factors (F(14, 4776)=12.23, p<.001) showed significant effects.

Figure 4.6 (left) shows that the Japanese group is sensitive to F0 fall in prominence perception, while in figure 4.6 (right), all the lines overlap and

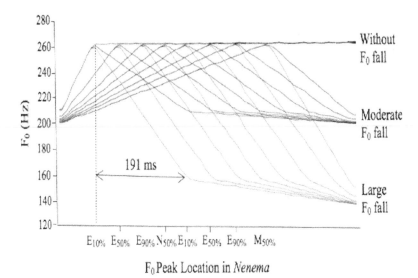

Figure 4.5   F0 Peak Location in Nenema. *Source*: Shport (2015: 20), reprinted with permission.

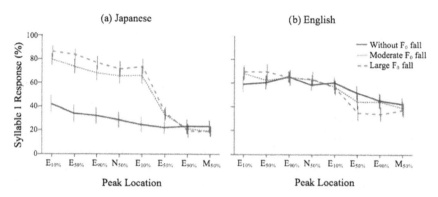

Figure 4.6   Syllable 1 Prominent Responses as a Function of F0 Peak Location and the Magnitude of the F0 Fall. *Source*: Shport (2015: 20), reprinted with permission.

we can see that English listeners are not sensitive to F0 fall. In the categorization task, the native language bias toward F0 fall is apparent, and Shport's prediction is again borne out.

Shport (2015) shows that English and Japanese listeners are both sensitive to F0 fall, but only Japanese listeners use F0 fall, that is, pitch movement, as a cue for prominence perception. Her observation is compatible with Sugito's idea of "late fall"; the F0 peak is identified as prominent even after it is produced, as long as it is accompanied by the F0 fall in Japanese.

In this section, we have studied prominence perception based on experiments with materials without lexical accents. We have argued that prominence in Japanese is cued not only by F0 peak but also by pitch movement. The prominence cues in Japanese are different from the ones in English. English listeners rely mainly on acoustic cues and use F0 peak and other segmental acoustic features of vowel duration and intensity to perceive prominence. A language probably has its own bias toward prominence perception. Our answer to the question "Why is L1 not easy to hear?" is that acoustic cues like F0 peak are perceived without much difficulty even by non-L1 listeners, while cues by pitch movements like F0 fall are difficult to perceive even for L1 listeners. Recall #7 in our perception experiment makes perceiving its prominence difficult. This is because #7 is not the F0 peak in φ_3; its phrasal H is lower than the following F0 peak H due to the pitch movement specified for φ_3. The notion of a language bias to perception might explain why L1 Dutch listeners show a lower focus identification rate than L1 Japanese and L1 Chinese listeners (cf. tables 4.2–4.3). Japanese and Chinese listeners are sensitive to F0-fall in their native languages, but Dutch listeners are not. We speculate Dutch listeners, like English listeners, use F0 peak location over F0 fall as a cue to prominence, and they do not perceive focal prominence well where the F0 peak and the focus location mismatch. We need further research to confirm this speculation.

We have another problem unsolved in this chapter. We do not see why focus identification is far better at #1 and #4 than #7. F0 peak and the focus location do not match in all of these positions, but the focus identification rates are different. The only difference is that the HL at φ_1 and φ_2 are followed by the boundary tone H, while at φ_3 it is not (cf. (4)). The boundary tone H might be effective to perceive F0 fall. We also need further research to solve this problem.

In this chapter, we come to know that Japanese has a language-specific perception bias and uses acoustic cues and pitch movement. In chapter 5, we will conduct a processing experiment on Japanese and English corrective focus to see whether the difference in perception cues affects language processing.

NOTES

1. Lee et al. (2019) later conducted perception experiments on Tokyo Japanese. We conducted our experiments before Lee et al. (2019). They claim that focus prosody of telephone numbers is not clearly marked and the identification rates are low in Tokyo Japanese. They analyze the results on acoustics only. We will discuss that how corrective focus is perceived cannot be accounted for by an acoustic analysis only in this chapter.

2. Japanese can drop subject, as far as it is recoverable from the context. We put *I* in the English gloss.

3. As observed in chapter 1, the terms of prosodic phrases vary among researchers. To avoid irrelevant confusion, we use the term φ-phrase in this chapter.

4. We see a difference between Lee et al.'s production data in table 4.1 and our materials in table 4.5. Our reviewer to the earlier draft of this book cast doubt to conduct a perception experiment by the material recorded by a single person. Lee et al. recruited five American English speakers and three Tokyo Japanese speakers, and we cannot see why the difference came from. We will discuss how our recorded materials are perceived by Japanese listeners and non-Japanese listeners in this chapter.

5. Minor φ-phrase is a near-equivalent to Accentual Phrase in the framework of Pierrehumbert and Beckman (1988). For the distinction between major and minor φ phrases, see Ito and Mester (2016).

6. Figure 4.3 can be found on the book's webpage on rowman.com.

7. Lee et al. (2019) consider the F0 peak followed by a fall as "lexical accent." We do not accept their idea, since the numbers in the 10-digit telephone sequence are devoid of lexical accents. We describe the F0 peak with a fall as HL here, instead of the H*L by Lee et al., to make it clear that the F0 peak with a fall is not a lexical accent.

Chapter 5

Neurocognitive Processing of Prominence

In chapter 4, we conducted perception experiments of corrective focus and found that L1 listeners and L2 listeners show similar perception tendencies, as summarized in table 5.1.

Gandour et al. (2007) is a classical study to investigate the neural substrates underlying the perception of prosodic phenomena. They conducted an fMRI experiment on 10 (M5, F5) late-onset intermediate Chinese-English bilinguals (average TOEFL score = 600, mean age = 27 (SD=3)) for sentence focus, that is, broad focus (BF) and contrastive focus (CF), and sentence type, that is, statement and question, discrimination tasks. Their results were that listeners earned a higher CF identification rate in Chinese, that is, their L1 (98.5%) than in English, that is, their L2 (86.5%). Also, their response time was longer in English (670.7ms) than in Chinese (552.4ms). They found that L1 and L2 elicited highly overlapping activations in the frontal, temporal, and parietal lobes. They also found that greater activation in L2 than in L1 occurred in the bilateral anterior insula and superior frontal sulcus. They conclude that L1 and L2 are medicated by a unitary neural system despite the age of acquisition and that additional neural resources may be required for unequal bilinguals.

Our perception experiment on corrective focus in chapter 4, however, had a result opposite to Gandour et al.'s data; L1 (Japanese) is harder to perceive corrective focus than L2 (English). Gandour et al.'s analysis would not account for our data. Japanese corrective focus is hard to perceive for L1 listeners as well as L2 listeners (cf. table 5.1). This means that the task difficulty measured by the focus identification rate is not brought by the L1/L2 distinction but by something else.

Hickok and Poeppel (2000) claim that similar sensory inputs involve a common neural system in the processing stream. In other words, different neural

Table 5.1   Identification Rates of Corrective Focus of Our Perception Experiments

	L1 listeners	*L2 listeners*
English	97.3%	98.6% (L1 Japanese)
Japanese	86.2%	85.5% (L1 Dutch)
		90.5% (L1 Chinese)

substrates are involved in different sensory inputs. Chinese and English probably have similar sensory inputs. Recall table 4.1 in chapter 4. Both Chinese and English cue corrective focus by acoustic features of duration, intensity, and F0. The focus identification rate is higher in Chinese (L1) than in English (L2). This is why greater activation is required to process English (L2), as Gandour et al. claim.

The story is different in Japanese/English bilinguals. We have found in chapter 4 that English and Japanese do not have similar cues to perceive corrective focus. F0 peak is an effective cue for English perception, while not only F0 peak but also F0 fall are effective in Japanese (cf. Shport, 2015). Based on Hickok and Poeppel (2000), we suspect Japanese and English involve different neural systems in processing corrective focus. Also, if the task difficulty brings graded activation in processing, as Gandour et al. (2007) claim, we would predict that Japanese requires more activation than English.

We will conduct an fMRI experiment below to see whether these predictions would be borne out.

5.1 PROCESSING OF CORRECTIVE FOCUS
IN JAPANESE AND ENGLISH

5.1.1 Methods and Materials

We recruited 22 right-handed late-onset Japanese/English bilinguals (M11, F11, mean age 26.7, SD = 11.1, average PBT TOEFL score = 595, SD = 55.7). They reported no auditory problems or a history of neurological or psychiatric disorders. All participants gave written consent before participating in the experiment. The experiment was approved by the Ethics Committee of Kobe University.

We used 80 tokens of English and Japanese broad focus[1] and corrective focus with 10-digit numbers (20X2 (broad/corrective) X2 (Japanese/English)), which are the same as the stimuli we used in the perception experiments in chapter 4. Before the experiment, participants received instruction about the distinction between broad and corrective focus in Question-Answer dialogues, like (1) and (2).

(1) English materials

    a. Corrective Focus (CF) condition   Q: Is Mary's number 215-418-5623?

                                            A: No, the number is 215-417-5623.

    b. Broad Focus (BF) condition   Q: What is Mary's number?

                                            A: The number is 215-417-5623.

(2) Japanese material

    a. Corrective Focus condition

        Q: Yamada-san-no   denwa-bango-wa   215-413-5623 desu-ka?

            Yamada-Mr/Ms-GEN phone-number-TOP   COPULA-Q

            "Is Yamada's number 215-413-5623?"

        A: Iie.   215-413-6623 desu.

            no.   COPULA

            "No. (It) is 215-413-6623."

    b. Broad focus Whole Sentence condition

        Q: Yamada-san-no   denwa-bango-wa   nan-ban

            Yamada-Mr/Ms-GEN phone-number-TOP   what-number

            desu-ka?

            COPULA-Q

            "What is Yamada's number?"

        A: 215-413-6623 desu.

                    COPULA

            "(It) is 215-423-6623."

Participants hear only the answer part, and judge whether or not the audio stimuli contained corrective focus. The "Yes" and "No" responses were provided by pressing the button with the index and the middle fingers of the left hand. They sat for a practice session before the experiment. The experiment was completed in about 1.5 hours, including the experiment explanation, consent credit, and post-experiment questions.

The fMRI paradigm we used was a 3T MRI scanner (Siemens, Prisma), Presentation (Neuro Behavioral System), and an MRI-compatible electrostatic headphone. We employed a block design with an average trial duration of 4 seconds and a response interval of 2 seconds. We had two experimental blocks (J/E order for 11 participants, and E/J order for the other 11). Data processing was carried out by Matlab 9.3 and analyzed by SPM 12 (Mathworks). Data were realigned to the first volume and normalized into standard stereotactic space (voxel size 3X3 mm, template provided by the Montreal Neurological Institute (MNI)).

5.1.2 Results

(3) gives behavioral results. The participants were more accurate in perceiving BF than CF both in English and Japanese.

(3) a. Accuracy (%):   English BF 95.19% (SD = 0.05) < Japanese BF
                       99.09% (SD = 0.01)

                       English CF 90.06% (SD = 0.11) > Japanese CF
                       80.02% (SD = 0.19)

   b. Response Time (ms):   English   451.6 ms < Japanese 467.3 ms

The accuracy of identifying CF is significantly lower in Japanese than in English (t (38) = −1.78, p = 0.042). The response time is shorter in English than in Japanese, but the difference is not statistically significant.

In chapter 4, we observed that the perception accuracy of Japanese corrective focus is 86.2%, and that of English corrective focus is 98.6% in our perception experiment. The accuracy rates are a bit lower in our fMRI experiment, but the identification tendency is almost the same; if we compare table 4.2 in chapter 4 with figure 5.1, we can see that out of the 10-digit number sequence, position #7 is a burden for listeners in both experiments.

Note that in our fMRI experiment, Japanese (L1) is worse in focus identification rate and longer in response time than English (L2). The behavioral data

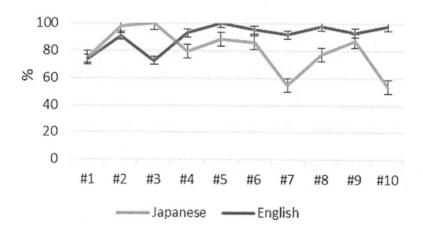

Figure 5.1   Corrective Focus Identification Rates Per Position of Ten-Digit Number Sequence.

of our experiment is incompatible with Gandour et al. (2007); in their experiment, Chinese (L1) is higher in focus identification rate and shorter in response time than English (L2). This difference leads us to suspect that Japanese, not Chinese or English, burdens listeners to perceive focus, and activates more neural areas.

Table 5.2 shows the activated brain regions specified with MNI when broad focus (above) and corrective focus (below) are processed in Japanese. Table 5.3 gives each value of the activated regions in MNI to process broad focus (above) and corrective focus (below) in English.

Tables 5.2–5.3 show that the regions responsible for auditory processing, that is, STG, SMG, Insula, are involved when corrective focus and broad focus are processed both in Japanese and English. Also, Pre-Motor areas are activated in both languages. What is different between the two languages is that as clearly observed in figure 5.2, the right Dorsolateral Prefrontal Cortex (DLPFC) is involved to process Japanese corrective focus, but not English corrective focus. This area is considered to be responsible for intonation processing (cf. Chien et al., 2020) and pitch memory (cf. Schaal et al., 2017) in the literature. Since our experiment materials are all statements and the intonational pattern is the same, we consider that pitch is involved to activate the right DLPFC in our experiment.

Table 5.2 Activated Areas (MNI) to Process Japanese Broad Focus (Above) and Corrective Focus (Below) (Random-Effect Group Analysis, $p < 0.05$, FEW Corrected)

PFEW-corr	Extent	Z-score	x	y	z	H	Anatomical Area	BA
0	6034	7.74	54	−22	4	R	PAC	41
0	5879	7.33	−52	−36	10	L	STG	22
0	2324	6.92	−4	16	44	L	Frontal Eye Field	8
0	737	6.06	38	−22	54	R	Primary Motor Cortex	4
0	321	5.68	−48	−34	46	L	SMG	40
0.001	227	5.6	56	4	44	R	Pre-Motor/SMA	6
0.001	95	5.59	−30	−2	48	L	Pre-Motor/SMA	6
0.005	68	5.28	−4	−22	30	L	Subcentral area	23
0.006	114	5.25	10	−20	4	R	Thalamus	
PFEW-corr	Extent	Z-score	x	y	z	H	Anatomical Area	BA
0	4820	7.61	54	−20	2	R	PAC	41
0	4439	7.41	−56	4	22	L	Pre-Motor/SMA	6
0	1501	7.12	40	−24	56	R	Primary Motor Cortex	4
0	2620	7.03	−6	4	58	L	Pre-Motor/SMA	6
0	319	5.82	−50	−28	44	L	SMG	40
0.001	60	5.78	32	40	26	R	Dorsolateral Prefrontal Cortex	9
0.001	156	5.63	54	4	42	R	Pre-Motor/SMA	6

SMG (Supramarginal Gyrus), STG (Superior Temporal Gyrus), SMA (Supplementary Motor Area), PAC (Primary Auditory Cortex), BA (Broadman Area), R (right hemisphere), L (left hemisphere)

Table 5.3 Activated Areas (MNI) to Process English Broad Focus (Above) and Corrective Focus (Below) (Random-Effect Group Analysis, p < 0.05, FEW Corrected)

PFEW-corr	Extent	Z-score	x	y	z	Hem.	Anatomical Area	BA
0	20730	7.84	62	−22	2	R	STG	22
0	3169	7.44	2	20	42	R	Frontal Eye Field	8
0	851	6.83	38	−48	44	R	SMG	40
0	1021	6.82	−36	−52	48	L	SMG	40
0.001	100	5.76	10	−66	46	R	Somatosensory Association Cortex	7
0.012	22	5.72	−4	2	30	L	Ventral Anterior Cingulate cortex	24

PFEW-corr	Extent	Z-score	x	y	z	Hem.	Anatomical Area	BA
0	16031	inf	−52	−32	10	L	STG	22
0	2903	7.28	10	20	40	R	Frontal Eye Field	8
0	689	6.31	−44	4	38	L	Pre-Motor/SMA	6
0	161	5.83	36	−50	44	R	Parahippocampal Cortex	36
0.002	493	5.52	40	−24	60	R	Primary Motor Cortex	4
0.002	228	5.51	−38	−46	42	L	SMG	40
0.004	40	5.39	30	48	2	R	Frontpolar area	10
0.003	49	5.38	10	−70	44	R	Somatosensory Association Cortex	7
0.005	78	5.28	34	−8	56	R	Pre-Motor/SMA	6

Table 5.4 is the result of the random-effect group analysis of English corrective focus vs. broad focus and Japanese corrective focus vs. broad focus. It shows that graded activities in left STG and SMA are observed to process corrective focus more than broad focus in English, but Japanese demands no such activation.

To see which language needs more activation to process corrective focus, we did a random-effect group analysis of English (CF-BF)–Japanese (CF-BF). Table 5.5 shows the result: bilateral STG and right PAC are more activated to process English focus than Japanese focus.

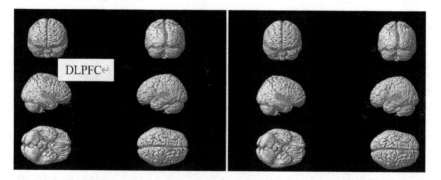

Figure 5.2 Activated Areas to Process Corrective Focus in Japanese (Left) and English (Right).

Table 5.4   Random-Effect Group Analysis of English (CF-BF) (Above) and Japanese (CF-BF) (Below) in MNI (Random-Effect Group Analysis, p < 0.05, FEW Corrected)

PFEW-corr	Extent	Z-score	x	y	z	Hem	Anatomical Area	BA
0.001	19	5.26	−52	−4	42	L	Pre-Motor/SMA	6
0.000	52	5.24	−54	−38	10	L	STG	22
0.017	8	4.93	58	−6	−2	R	PAC	41

PFEW-corr	Extent	Z-score	x	y	z	Hem	Anatomical Area	BA
0.006	8	4.96	−38	−22	60	L	Primary Motor	4
0.022	2	4.83	−42	−16	60	L	Pre-Motor/SMA	6

5.2 DISCUSSION

Our research questions (RQ) were (i) Do Japanese and English involve different neural systems in processing corrective focus? (ii) Does Japanese require more activation than English to process corrective focus? Our answer to RQ (i) is "yes"; Japanese activates the auditory-motor interface system and English uses the left temporal-parietal-occipital junction (cf. figure 5.2 and table 5.5).

Green and Abutalebi (2013) explore the nature of the control process in the neural network in bilingual speakers and propose the Adaptive Control Hypothesis. They claim that language control processes adapt to the recurrent demand placed on them by the interactional context and change a parameter or parameters about the way it works. Figure 5.3 provides a schematic description of the neural structures and the connections they associate with language control processes. They identify ACC and pre-SMA with conflict monitoring, and pre-SMA initiates speech in language switching. They associate left PFC with the control of interference, and parietal cortices with the maintenance of task presentation. Basal ganglia switch languages.

Applying figure 5.3 to our study, we can claim that Japanese/English bilinguals activate motor areas to switch languages and change the parameters in their neural network. Tables 5.2–5.3 illustrate that left STG and INS are employed to process both English and Japanese corrective focus, but Japanese

Table 5.5   Random-Effect Group Analysis of English (CF-BF) vs. Japanese (CF-BF) in MNI (Random-Effect Group Analysis, p < 0.05, FEW Corrected)

Extent	Z-score	x	y	z	Anatomical Area	BA
410	5.49	−54	−14	0	STG	22
45	5.04	60	−18	4	PAC	41
3	5.01	−50	−2	40	Pre-Motor/ SMA	6
13	5.01	62	−26	10	PAC	41
5	4.99	52	−6	−2	STG	22

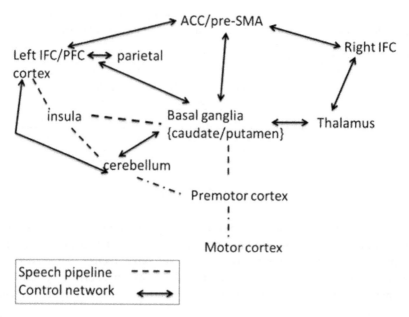

Figure 5.3 **Language Control Network and Speech Production Regions.** *Source*: Green and Abutalebi (2013: 523), reprinted with permission.

also needs to activate the right DLPFC and Thalamus. We see that Japanese/ English bilinguals switch parameters, depending on the language they use.

Perrone-Bertolloti et al. (2013) investigated the perception of broad focus and corrective focus in French, and report that bilateral Inferior Frontal Cortex (IFC) and SMG, left STG, Middle Temporal Gyrus (MTG), pre-Motor, and Insula (INS) are more activated to process corrective focus in French. If we compare the contrast between CF and BF both in Japanese and in English, table 5.4 shows that graded activities in left STG and SMA are observed to process CF more than BF in English, but Japanese demands no such activation.

Our answer to RQ (i) "Do Japanese and English involve different neural systems in processing corrective focus?" is that Japanese and English use different neural systems to process corrective focus.

Our RQ (ii) is "Does Japanese require more activation than English to process corrective focus?" Table 5.5 shows that English focus requires more activation than Japanese focus. English, an L2 of our participants, demands more neural activation than Japanese, the L1 of our participants. Our data is incompatible with the Chinese/English bilinguals' behavior reported by Gandour et al. (2007). In their experiment, L2 (English) is more difficult to process than L1 (Chinese) and L2 is lower in focus identification rate and longer in response time of focus. They claim that the activity in aINS (anterior insular) is graded in response to the task difficulty in L2.

Their data, however, can be differently analyzed in the Adaptive Control Hypothesis by Green and Abutalebi (2013). If Chinese/English bilinguals demand more activation in aINS, they would use the same parameter to process Chinese and English. Recall that Chinese and English use acoustic cues of duration, intensity, and F0 for focus identification (cf. figure 4.1 in chapter 4). They have similar sensory inputs, and as Hickok and Poeppel (2000) claim, use a common neural system in the processing stream. It is a natural recourse that the task difficulty in L2 requires more activation in aINS than L1.

The story is different for Japanese/English bilinguals. They switch parameters to process Japanese and English. If neural substrates are different between Japanese and English, it does not matter which language demands more neural activation. We conclude that Japanese and English use different neural substrates due to the difference in sensory inputs. The task difficulty, measured by focus identification rate and response time, is not brought by the L1/L2 distinction but by the difference in sensory inputs.

Before we close this chapter, we would like to reconsider the "broad focus" in Japanese. In our fMRI experiment, we asked our participants to distinguish "corrective focus" and the focus on the whole sentence. As we discussed in chapter 2, Halliday (1967) distinguishes "broad focus" from "narrow focus." He defines the focus on the whole sentence as "broad focus." Broad focus highlights the non-focal phrase-final element, due to Nuclear Stress Rule (cf. Chomsky and Halle, 1968) in English. "Corrective focus" realizes focal prominence and narrows the domain of a focus. English distinguishes broad focus and narrow focus, and as table 5.4 shows, corrective focus demands more neural activation than broad focus. Japanese, on the other hand, hardly realizes a difference in the activation of the neural substrates in processing corrective focus and broad focus (cf. table 5.4). There might be two possible explanations for this phenomenon. One is that the Japanese distinction between corrective focus and broad focus does not generate enough difference in sensory input to lead to a difference in neural activation. The second is that Japanese does not have a broad focus. We claim that the second possibility is more correct. If Japanese does not mark the focus on the whole sentence, the results in table 5.4 looks very natural. Our idea is opposite to Ishihara (2003) and Ishihara et al. (2018). They discuss that Japanese has non-focal prominence, assigned to the immediately preverbal phrase, evidenced by a production and perception experiment on the sentence level. Prominence is a perceptual feature, and the authors personally do not share the perceptual intuition with them. We will conduct a perception experiment on spontaneous

Japanese in chapter 6 and try to provide quantitative confirmation of focal and non-focal prominence on the utterance level.

NOTE

1. In chapter 2, we discussed that Japanese probably does not realize broad focus prosodically, contra English. We, however, use the term "broad focus" to refer to the focus of the whole sentence, at the cost of possible confusion.

Chapter 6

Prominence in Spontaneous Japanese

Prominence is a speaker's intuitive sensation of strength in an utterance he or she hears (cf. Büring, 2016: 1). Traditional studies on prosody argue that prominence is highly tied to changes of F0. The literature (cf. Ladd et al., 1994 for English, Rietveld and Gussenhoven, 1985 for Dutch, among many others) reports that segmental acoustic properties are important prominence perception cues in English and other stress-accented languages. Japanese is a mora-timed pitch language, and it is not a surprise that it employs different cues from stress-accented languages in prominence production and perception.

The literature classifies two types of prominence: alignment/non-focal prominence and focal prominence. The former aligns prominence syntactically and phonologically, and the latter aligns prominence on focus. Languages vary in which prominence they mark. Based on production and perception experiments, we have concluded that Japanese does not mark alignment/non-focal focus. It only marks focal prominence. The literature has analyzed many types of focus (cf. Büring, 2016; Wagner, 2020 for the list of focus) and we have so far recognized that Japanese marks information-new focus and corrective focus in chapters 3 and 4. We studied how the focal prominence on the sentence level is produced by a speaker and perceived by a listener in the previous chapters and found that phonetic cues called "focus effects" are not as strong as claimed in Japanese literature. Instead, prosodic pitch movements are important production cues, and the locus of prominence in a prosodic phrase also plays an important role in perception.

We further studied the processing of focal prominence in chapter 5 and found that Japanese uses different substrates from English to process corrective focus, and we have provided a piece of evidence that the sensory input of prominence is different between Japanese and English.

Our findings so far are on the prominences on the sentence level, and we still do not know how prominence is produced and perceived on the utterance level. In this chapter, we will conduct a perception experiment of prominence on the utterance level. The methodology we use is Rapid Prosody Transcription (RPT). Recall that prosodic boundaries have not been established in Japanese literature (cf. chapter 1). We will work on how boundaries and prominences are marked and perceived in Japanese below.

6.1 METHODOLOGY: RAPID PROSODY TRANSCRIPTION (RPT)

Rapid Prosody Transcription is a challenge to broaden a research target to cover spontaneous speech, which is originally developed by Cole and her colleagues (cf. Cole et al., 2010a, 2010b, among others). They consider that listeners differ in how they perceive prosody for the same utterance. They suspected acoustic, phonological, syntactic, semantic, and pragmatic properties determine how prosody is perceived. They used prosodic transcription as a tool for prosody research. Listeners identify prominences and boundaries based on an auditory impression of an utterance. Transcribers listen to recorded speeches up to 25 seconds of 38 short excerpts from Buckeye Corpus (Pitt et al., 2007) through headphones or speakers and are asked to mark prominences by an underline and boundaries by a vertical line on individual words in a transcript with all punctuation and capitalization removed, as in (1).

(1)   . . . i really don't know i think in today's world what they call the nineties that uh it's just like everything is changed like when i grew up . . .

Transcribers are given minimal instructions such as "mark prominent words that the speaker has highlighted for the listener, to make them stand out" and "mark boundaries between words that belong to different chunks that serve to group words in a way that helps listeners interpret the utterance." They are given no example transcriptions and no feedback on their transcription.

RTP can be used to gauge the perceptual salience of prominences and boundaries in speech representing any genre, style, or language. Inter-transcriber variability is used to calculate continuous-valued prosody "scores" that are assigned to each word and represent the perceptual salience of its prosodic features or structure. The boundary (b-) score and the prominence (p-) score ranging from 0 to 1 are calculated, as illustrated in table 6.1.

Table 6.1 Prominences and Boundaries Coded

				Boundary				Prominence			
Token	Word	B-Score	P-Score	A1	A2	A3	A4	A1	A2	A3	A4
1	i	0	0	0	0	0	0	0	0	0	0
2	really	0	0.56	0	0	0	0	0	1	1	0
3	don't	0	0.25	0	0	0	0	0	0	0	0
4	know	0.81	0.44	1	1	1	0	0	0	0	0
5	i	0	0	0	0	0	0	0	0	0	0
6	think	0.31	0.44	0	0	0	0	0	1	0	0
7	in	0	0	0	0	0	0	0	0	0	0
8	today's	0	0.81	0	0	0	0	1	1	0	0
9	world	0.81	0.44	1	1	1	0	0	0	0	0
10	what	0	0	0	0	0	0	0	0	0	0
11	they	0	0	0	0	0	0	0	0	0	0
12	call	0	0	0	0	0	0	0	0	0	0
13	the	0	0	0	0	0	0	0	0	0	0
14	nineties	0.81	0.5	1	1	1	0	1	0	0	1
15	that	0.5	0.13	0	1	1	0	0	0	0	0
16	uh	0.38	0	1	0	0	0	0	0	0	0
17	it's	0	0	0	0	0	0	0	0	0	0
19	like	0	0.06	0	0	0	0	0	0	0	0

B-scores and p-scores indicate the proportion of participants who under-score the respective word, and higher values indicate strong perceptual salience of the prosodic element (for further details, see Cole and Shattuck-Hufnagel, 2016). RPT has no restriction on its materials or subjects, and the previous studies employ spontaneous speech or read speech, L1 or L2 listeners.

To assess inter-transcriber agreement, Cole and her colleagues used Fleiss' Kappa, and they reported that agreement between transcribers is highly significant for labeling of prominent words (mean Kappa = 0.392) and boundaries (mean Kappa = 0.582) in their experiments on English spontaneous speech (cf. Cole et al., 2010a).

For syntactic analysis of boundary marking, they coded each word for the highest syntactic category at its left and right edges, following Penn Treebank guidelines, as in (2). (3) lists the syntactic and non-syntactic categories they consider, to analyze the experiment materials.

(2) $[_S$ I $]_{NP}$ $[_{VP}$ think $]_{w/p}$ $[_{PP}$ in today's world . . .

(3) syntactic coding

  S: matrix S

  S-bar: subordinate or relative clause

  S2:   S preceded by conjunction or relative pronoun

CC-S: coordinating conjunction preceding or following a sentence

CC-XP   coordinating conjunction preceding or following an XP

Phrase (XP):   any XP that is not a clause

Within Phrase (w/p):   any word boundary that does not align with a coded syntactic boundary

Disfluency: filled pause, repetition, and repair disfluencies

Discourse Marker (DM):   *yknow, like, so, I mean*, etc.

Cole et al. (2010a) found that boundary scores decrease at the right edge of syntactic units from higher-level units (clauses) to lower-level ones (XP, word) in English (cf. figure 7.2 in chapter 7).

Wightman et al. (1992) claim that final vowel lengthening effects are robust cues to prosodic boundaries in English and Cole et al. (2010a) investigated whether vowel duration directly encodes the syntactic category. They found a weak correlation between syntactic categories and vowel duration in English. They conclude that the first predictor of boundaries is syntactic context, and the second predictor is vowel duration in English.

Prominence is more complex than boundaries to analyze. Cole et al. (2019) is a comparative study of English, French, and Spanish prominences, and they report that prominence ratings are statistically different ($\chi^2(2) = 12.18$, $p = 0.002$) among the three languages. They claim that the three languages are similar in the use of acoustic/phonetic cues, but they vary in how non-phonetic cues like usage frequency, accent, and ToBI pitch accent affect the prominence perception.

6.2 RPT EXPERIMENT ON SPONTANEOUS JAPANESE

6.2.1 Method and Materials

We conducted RTP experiments online via Yahoo! Crowd-Sourcing service. Our RPT experiment was approved by the Ethical Committee of the National Institute of the Japanese Language and Linguistics (NINJAL). We used 12 materials of *Corpus of Spontaneous Japanese* (CSJ), released by NINJAL. CSJ is a corpus of pseudo-lectures and monologues collected from adult Tokyo Japanese speakers and we chose 12 excerpts, that is, 6 excerpts each.[1] Since Japanese is a head-final agglutinative language, we segment our data set on the morpheme level, as in (4).

(4) $[_{ADV}$ mazu $]_{ADV}$ $[_{NP}$ $[_{S2}$ $[_{N}$ watashi$]_{w/p}$ - $[_{PART}$ no$]_{w/p}$ $]_{w/p}$ $[_{VP}$ sukina$]_{w/p}$ $]_{S2}$
first   I   GEN   like

$[_{N}$ mono$]_{N}$ $]$ $_{NP}$ $[_{VP}$ $[_{N}$ inu$]_{w/p}$ $-$ $[_{AUX}$ desu$]_{AUX}]$ $_{VP}$
thing   dog   COPULA

"First, what I like is a dog."

(N.B. In the experiment, the text is written in Japanese *Hiragana*, that is, Japanese cursive character, without punctuation.)

Generally, the syntactic category of spontaneous speech is coded on the word level in RPT. Baumann and Winter (2018) argue that prominence is marked only on content words, not on function words in German. But as we have already observed in chapter 1 (cf. figures 1.2 –1.3), Japanese marks function morphemes such as particles and auxiliaries prominent. We believe that we need a syntactic coding system different from English proposed by Cole et al. (2010a) and decided to code our material on a morpheme-based parsing. Our material (see appendix E) contains 980 morphemes in total, and table 6.2 gives the token distribution of each part of speech.

Among the 980 morphemes, 232 are accented and 339 are unaccented, and the rest is function morphemes with no lexically given accents (cf. (10) in chapter 2 for lexical accent in Japanese, and figure 1.1 in chapter 1 for the pitch movement of lexical sequences). Our materials were 18- to 40-second long in duration.

Table 6.2 Token Distribution of Part-of-Speech Categories (Morpheme-Based)

Content Words		*Function Morphemes*	
Noun	264	Particles	177 (case particles like *ga* (SUBJ), *o* (OBJ), *ni* (Loc), etc.)
Verb	137	Auxiliary Verb	89
Conjunction (*de* "and")	37	Discourse Marker	60 (*ano* "the year," *e-to* "eh," etc.)
Adverb	27	Suffix	41 (*ha* "clan," *o-* (honorific), etc.)
Adjective	22	Complementizer	31 (*to, no*, etc.)
WH	2	Numeral	27
Pronoun	1	Topic Marker	25 (Topic construction: x-*wa* . . .)
		Demonstrative	16
		Classifier	6
		Negation	3
		Quantifier	1
		Disfluency	10
		Interjection	4

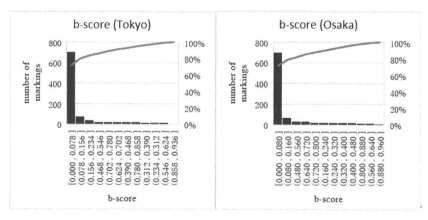

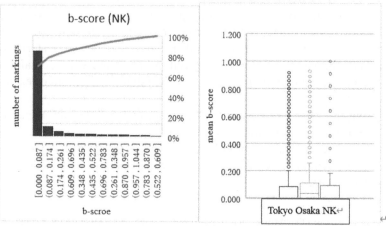

Figure 6.1   Histogram of b(oundary)-Scores.

We recruited three groups of listeners: 35 Tokyo Japanese listeners (mean age 24.8, SD=0.5), 27 Osaka Japanese listeners (mean age 25.3, SD=3), and 11 Northern Kanto (NK) Japanese listeners (mean age 23.2, SD=4)[2] to see dialectal differences among the listeners. Japanese has many dialects whose prosodic systems are so different from each other that they sound "as if they are different languages (cf. Sugito, 2001: 197)." The three dialects we have chosen are different in lexical accent system, dephrasing, and tone scaling; Tokyo and Osaka Japanese are accented dialects, and the accent system is more complex in Osaka Japanese. NK Japanese is an accentless dialect, that is, it does not mark H*+L on the word level. When two words are combined to form a phrase, the phrase at the word level is deleted. It is called "dephrasing" in Japanese literature. Whether or not a dialect applies dephrasing leads to a difference in pitch movement. Tokyo Japanese has dephrasing, while

Osaka Japanese does not. In other words, Osaka Japanese keeps a lexical accent on the upper phrase level above the word level (cf. Igarashi, 2014). On the utterance level, Japanese forms the downward tone scaling in Tokyo Japanese, the upward tone scaling in NK Japanese (cf. Igarashi, 2014; Igarashi and Koiso, 2012, among others), and mixed tone scaling in Osaka Japanese (cf. figure 7.3 in chapter 7). These are the general differences in production, and we wonder if they would affect boundary and prominence perception. Our participants had no hearing difficulties. After the exercise session, they listened to each material twice via PC, marking prominences and boundaries by clicking a mouse simultaneously. Their responses were saved on the computer via LMEDS, developed by Mahrt (2016). It took about 30 minutes for our participants to complete the task. They were paid by Yahoo! points.

6.2.2 Results

The overall inter-listener agreements were κ0.638 on the boundary (b-) score and κ0.359 on the prominence (p-)-score on Fleiss' Kappa. Comparing these figures with English ones in section 6.2.1, we believe that our scores are reliable above the chance level. The correlation between the p-score and b-score is weak (r = 0.12 on Person's Correlation). Figure 6.1 shows the histograms of b-scores marked by the listeners of the three groups.

One-way ANOVA between the b-score and three dialects was $F(2, 2940) = 0.48$, $p = 0.01$, and further Bonferroni is significant only between the Tokyo dialect and Osaka dialect (p = 0.01).

Table 6.3 shows the number of b-scores ($>0.2^3$) marked by the participants of each dialect, and figure 6.2 shows the mean b-scores of the major boundary categories at the right edge, that is, where a phrase boundary ends. As we will see later, the syntactic categories used as boundary cues are different in Japanese from English (cf. Cole et al., 2010a); both Japanese and English listeners use higher categories S and Conjunction as boundary cues, but Japanese listeners use lower XP categories as well (cf. figure 6.2, figure 7.2 in chapter 7).

Table 6.3   The Number of B(oundary)-Scores (>0.2) Marked by the Participants

	S	S2	SBAR	CP	TopP	NP	N	VP	ADVP
Tokyo	31	6	1	0	23	27	10	7	3
Osaka	31	6	1	1	24	31	10	7	3
NK	37	5	1	0	21	23	10	7	2

	ADV	ADJP	ConjP	Conj	W/P	DM	DISF	Suffix	Total
Tokyo	3	1	7	5	12	21	0	8	167
Osaka	4	1	7	5	12	27	1	8	181
NK	2	1	8	5	12	21	0	8	163

MEAN B-SCORE AT RIGHT EDGE

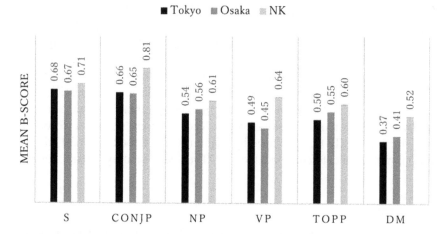

Figure 6.2 **Mean b-Scores at the Major Boundary Categories at the Right Edge.**

To see whether or not acoustic cues are effective for boundary perception, we conducted a regression analysis between b-scores and maximum F0 (MaxF0), range F0, duration, and intensity. The results are given in table 6.4.[4] Adjusted r^2s are very low, and we conclude that acoustic cues are not boundary predictors for the listeners of all three groups.

The story is different in prominence perception. One-way ANOVA between p-score and three dialects is $F(2, 2940) = 1.55$, $p < 0.001$, and further Bonferroni is significant in Tokyo X Osaka ($p = 0.001$), Tokyo X NK as well as Osaka X NK ($p < 0.001$, for both) (cf. figure 6.3).

Table 6.5 is the results of regression analyses between p-scores and acoustic cues in the three dialects we consider. We have found that the acoustic cues of MaxF0 and intensity are not strong predictors of prominence in Japanese. Range F0 is a moderate predictor and is more effective than duration in Japanese prominence perception.

Table 6.6 is the results of the regression analyses between p-scores and a position in a prosodic phrase. We can see that the phrase-initial position is

Table 6.4 **Results of Regression Analyses between B-score and Acoustic Cues**

	Tokyo Japanese		Osaka Japanese		NK Japanese	
	p	Adjusted r^2	*p*	Adjusted r^2	*p*	Adjusted r^2
b × MaxF0	0.33	< 0.001	0.911	0.001	0.59	0.0007
b × range F0	0.002	0.009	< 0.001	0.015	< 0.001	0.011
b × duration	< 0.001	0.036	< 0.001	0.056	< 0.001	0.041
b × intensity	0.67	–0.0008	0.93	–0.001	0.57	–0.0007

a better predictor of prominence, followed by the phrase-mid position in all three dialects under consideration.

To see a relation between the p-score and a position within a phrase, we have counted the number of prominences per boundary of our data set. Table 6.7 shows the numbers of prominence markings (p > 0.2) and boundary markings (b > 0.2) by the listeners of each dialect.

The three dialects under analysis vary in prominence marking more than in boundary marking. Comparing the numbers of p-marking and those of b-marking, we can see that the three groups differ in the frequency of p-marking per intonation phrase (IP). The number of prominences per IP varies from zero (cf. figure 6.4 left) to three (cf. figure 6.4 right) in our data set.

To sum up, we have found (i) prominence marking significantly varies among the listeners of the three dialects we consider (cf. table 6.7), (ii) prominence is not predicted by acoustic cues of maximum F0, duration, or

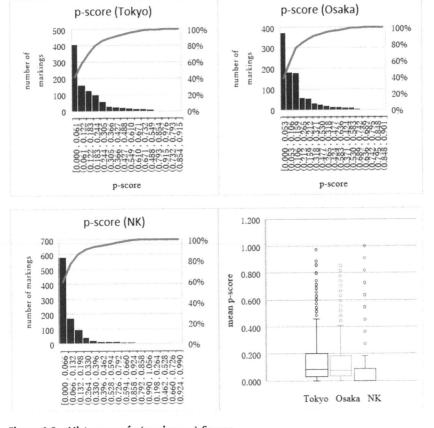

Figure 6.3   Histogram of p(rominence)-Scores.

Table 6.5   Results of Regression Analyses between P-scores and Acoustic-Cues

	Tokyo Japanese		Osaka Japanese		NK Japanese	
	p	Adjusted r²	p	Adjusted r²	p	Adjusted r²
p × MaxF0	< 0.001	0.159	< 0.001	0.172	< 0.001	0.121
p × range F0	< 0.001	0.237	< 0.001	0.252	< 0.001	0.199
p × duration	< 0.001	0.147	< 0.001	0.146	< 0.001	0.097
p × intensity	< 0.001	0.051	< 0.001	0.051	< 0.001	0.04

intensity, but range F0 is a moderate predictor of prominence (cf. table 6.5), (iii) not only content words but function morphemes are subject to get prominent in Japanese (cf. table 6.7, figure 4.4 (right)), (iv) prominence marking does not necessarily match boundary marking, and the number of perceived prominences varies in a prosodic boundary (cf. figure 4.4, table 6.7), and (v) as for boundary markings, acoustics does not work as a cue (cf. table 6.4) but syntactic categories are more reliable cues (cf. figure 6.2, table 6.3).

6.2.3 Discussion

We conducted RPT experiments to see how boundaries and prominences are marked and perceived in spontaneous Japanese. The results of our experiments in the previous section are given statistically, and we will observe tone movements below to account for what these stats mean.

Figure 6.5 is an example with six intonation phrases (IP), and table 6.8 is its coding of boundaries and prominences perceived by the listeners of the three dialects.

There is not much difference in boundary coding but we can see that prominence coding varies among the three dialects. In table 6.8 Token 5 *doshite* "why" is locally boosted in F0, as observed in figure 6.5 and its p-scores are relatively high in all the dialects. It is interesting that *doshite* is introduced into the utterance in Token 3, but the same expression *doshite* is locally boosted in Token 5 and is perceived more prominent than in Token 3. This is an example where given information is marked as prominent. Token 9 *supichi* "speech" and Token 11 *heta* "poor" are examples of local F0

Table 6.6   Results of Regression Analyses between P-scores and Position in a Phrase

	Tokyo Japanese		Osaka Japanese		NK Japanese	
	p	Adjusted r²	p	Adjusted r²	p	Adjusted r²
p × phrase-initial	< 0.001	0.287	< 0.001	0.29	< 0.001	0.328
p × phrase-mid	< 0.001	0.175	< 0.001	0.246	< 0.001	0.213
p × phrase-final	< 0.001	0.066	< 0.001	0.087	< 0.001	0.094

Table 6.7 Number of Prominence and Boundary Markings

	Prominence-marking (p > 0.2)				Boundary Marking (b > 0.2)
	Phrase-initial	*Phrase-mid*	*Phrase-final*	*total*	*Total*
Tokyo Japanese	74	145	53	272 (41)	167
Osaka Japanese	63	91	26	180 (21)	181
NK Japanese	52	54	22	128 (13)	163

(*Note:* The number in the parentheses is the number of *p*-markings on a function morpheme)

boost. The former is an accented content word and the latter is an unaccented content word. Token 15 *to* is a complementizer and it is assigned no lexical accent. It is assigned a boundary tone HL%, and is perceived as prominent by Osaka and NK listeners but not by Tokyo listeners. At the moment we have no explanation for the difference among the dialects. We will leave it for the future research.

Figure 6.6 is another example of a local boost of an accented word and an unaccented word as well as boundary tones on function morphemes that are perceived as prominent. In IP1, two content words *yuyo* "useful" in Token 1 and *konchu* "insect" in Token 2 are perceived as prominent and these words keep their lexical accents LH and H*+L.[5] *Toshite* "as" at the IP-final position is a function morpheme and is unaccented lexically. It is highlighted by a boundary tone HL% and is perceived as prominent by the listeners of all three dialects. IP1 has three prominences in total. IP2 and IP4 are not perceived as prominent. IP3 contains one content word Token 6 *tema* "theme," followed by particle *no* (GEN). *Tema* keeps its lexical accent H*+L and particle *no* (GEN) does not have a lexically determined accent. Here *no* is assigned a boundary tone H%. Both *tema* and *no* are perceived as prominent in IP4 by the Tokyo and Osaka listeners (cf. table 6.9).

What is noteworthy is that content words keep their lexical pitch accent, that is, H*+L. Function morphemes are either lexically accented, as in *kara* "from" and *made* "even," or not lexically accented, like *no* "GEN." One-mora

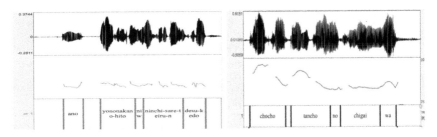

Figure 6.4 Pitch Movement with and without a Prominence Marked.

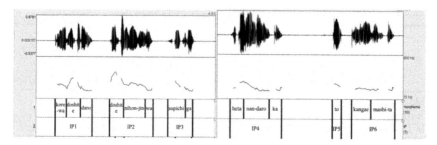

Figure 6.5   Pitch Movements across Multiple IPs. (Extracted from CSJ file #S00f0208)

function morphemes like *ga* (NOM), *o* (ACC), or *wa* (TOP) do not have a bitonal accent by definition. They are either attached to a preceding content word or are assigned a boundary tone, that is, H%, LH%, or HL% at the IP-final position. It is the latter, not the former, that is perceived as prominent. We can say that boundary tones H%, LH%, and HL% are prominence-lending tones.

To sum up, we have observed that there are two ways to mark prominence in Japanese: local F0 boost and boundary tone. The former is assigned locally to content words and the latter is assigned to function morphemes at

Table 6.8   Coding of Boundaries and Prominences in Figure 6.5[6]

Token	Morpheme	B(oundary)-score			P(rominence)-score		
		Toyo	Osaka	NK	Toyo	Osaka	NK
1	kore	0	0	0	0.14	0.11	0
2	wa	0	0	0	0.09	0.11	0
3	doshite	0	0	0	0.17	0.11	0.36
4	daro	0.71	0.78	0.82	0.14	0.15	0.09
5	doshite	0.06	0.07	0.09	0.71	0.48	0.55
6	nihon	0	0	0	0.29	0.19	0
7	jin	0	0	0	0.26	0.19	0
8	wa	0.57	0.7	0.64	0.23	0.15	0
9	supichi	0.03	0.04	0	0.31	0.37	0.27
10	ga	0.51	0.67	0.73	0.17	0.22	0
11	heta	0	0.04	0	0.4	0.44	0.27
12	nan	0	0	0	0.23	0.19	0.18
13	daro	0.03	0	0	0.26	0.19	0.18
14	ka	0.66	0.82	0.82	0.23	0.19	0.09
15	to	0.17	0.37	0.18	0.17	0.26	0.36
16	kangae	0	0.04	0	0.06	0.07	0
17	mashi	0	0	0	0	0	0
18	ta	0.17	0.11	0.36	0.06	0	0

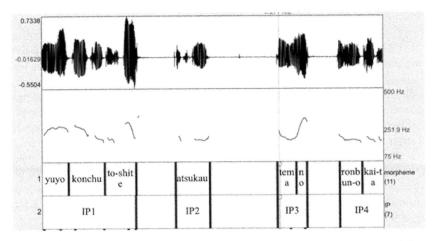

Figure 6.6   Prominence of Content Words and Function Morphemes. (Extracted from CSJ file #S00f0082)

the boundary final position. In Japanese literature, the function of boundary tones is analyzed as "emphasis" (cf. Kawakami, 1963) or "delimitative" (cf. Taniguchi and Maruyama, 2001; Maekawa, 2011). We do not think the delimitative function is a necessary or sufficient condition of boundaries. Recall figure 1.2 (b) in chapter 1; it marks H% without a boundary marking. Needless to say, most boundaries are without prominence-lending boundary tones of H%, LH%, or LHL%.

We will discuss what prominence does in Japanese and why Japanese has two ways of highlighting systems in chapter 7.

Table 6.9   Coding of Boundaries and Prominences in Figure 6.6

Token	Morpheme	B(oundary)-score			P(rominence)-score		
		Toyo	Osaka	NK	Toyo	Osaka	NK
1	yuyo	0	0.07	0	0.49	0.26	0.46
2	konchu	0	0.07	0.09	0.46	0.26	0.46
3	toshite	0.49	0.56	0.64	0.46	0.44	0.46
4	atsukau	0.6	0.7	0.73	0.23	0.07	0.09
5	tema	0	0.04	0.09	0.43	0.26	0.36
6	no	0.46	0.63	0.27	0.43	0.3	0.18
7	ronbun	0.03	0.04	0	0.14	0.07	0
8	o	0	0.07	0.09	0.06	0	0
9	kai	0	0.04	0	0.03	0.04	0
10	ta	0.14	0.19	0.27	0.03	0	0

NOTES

1. CSJ is a big corpus of monologues and dialogues by more than 1,400 native Japanese speakers. It contains 3,302 files, and the total recorded time is more than 660 hours (cf. Maekawa 211:13). The materials we used for our experiments are of very small size. We selected 13 excerpts including the exercise set without a textual bias and with clear vocal quality and enough pitch range to analyze the sound waves. We restricted ourselves not to exceed 30 minutes for a perception experiment, and the 13 excerpts in total were the maximum for that purpose.

2. We recruited 50 participants for each group. Due to the Covid-19 pandemic, we could not conduct our RPT experiments in person. In our online experiments, we asked our participants to claim their native dialect. We found a certain number of mismatches between a self-claimed native dialect and the target dialect, and it is our regret that we had to dismiss the mismatched data.

3. We have decided to choose b > 0.2 because it is a little bit above the average score.

4. The units of the four acoustic measures are not the same, so we calculated z-scores of each acoustic measure for the regression analyses.

5. *Konchu* "insect" is lexically unaccented, that is, LH, but when it is combined with an adjective, as in *yuyo konchu* "useful insect," the compound accent rule applies and *konchu* gets accented, that is, H*+L.

6. See Appendix E for the gloss of each token.

Chapter 7

What Does Prominence
Do in Japanese?

Japanese literature has long studied lexical pitch accents on the word level and phrasal tones on the clause level. Most of them are production studies and aim to account for how Japanese prosody is formed compositionally. What makes Japanese prosody look complicated is that both lexical accents and phrasal tones are realized by the H/L bitonal system. Lexical accents are determined at the word level, but they are subject to change to form a compound (cf. (1a)) or a phrase (cf. (1b)).

(1) a. ya' ma (LH*(+L)) + sakura (LHH) → yama'zakura (LHH*+LLL)
         mountain   cherry   mountain cherry
     b. kii'roi (LHH*+L) + 'yane (H*+L) → kiiroi 'yane (LHHHH*+L)
         yellow   roof   yellow roof

Ya 'ma "mountain" and *kii'roi* "yellow" are lexically accented in Tokyo Japanese. When they form a compound or a phrase, they get deaccented. (1) is just a simple example, but Japanese compound rules and phrase-making rules are far more complicated (cf. Kubozono, 1988; Akinaga (ed.), 2014, among many others). Japanese literature has boasted a tremendous amount of work on these issues, but we will not go into them here. Instead, we will focus on how tones on multiple prosodic phrases interact and form Japanese prosody compositionally.

Pierrehumbert and Beckman (P&B) (1988) posit that tone changes as observed in (1) occur at their Accentual Phrase (AP). As discussed in chapter 1, we do not use terms that are specific to Japanese literature but assume (2) as the universal prosodic hierarchy. A tone change like (1), that is, the change in lexical accent, occurs on a φ-phrase in our framework.

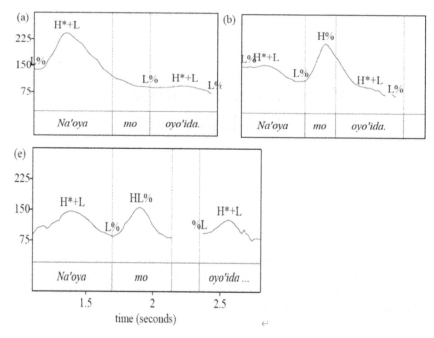

Figure 7.1   Pitch Movements of Na'oya-mo oyo'ida. Venditti et al. (2008: 488), reprinted with permission from the Oxford University Press.

(2)   Prosodic Hierarchy (cf. Féry, 2017: 36)

υ	Utterance	(corresponds roughly to a paragraph or more)
ɩ-phrase	intonation phrase	(corresponds roughly to a clause)
φ-phrase	prosodic phrase	(corresponds roughly to a syntactic phrase)
ω-word	prosodic word	(corresponds roughly to a grammatical word)
F	Foot	(metrical unit)
σ	syllable	(strings of segments)
μ	Mora	(unit of syllable weight)

The lexical accent is not the only tone observed in Japanese prosody. Recall figure 1.2 in chapter 1 (repeated as figure 7.1), where "boundary tones" L%, H%, and HL% are assigned. Boundary tones are the tones that are assigned at the beginning and the end of a prosodic boundary. Tokyo Japanese, for example, initiates and finalizes a statement with a low tone. The former is called Initial Low (%L) and the latter is a phrase-final falling tone (L%). Besides these low tones, other local pitch movements are observed at a phrase boundary; Venditti et al. (2008) list H%, HL%, LH%, and HLH% as Boundary Pitch Movements (BPM). These tones differ from lexical tones since they are not assigned to a specific lexical item. They appear at the phrase's final position, and they are considered to add semantic or pragmatic meaning. Maekawa (2011: 146) reports that about 25% of all the accented phrases are accompanied by BPM in *Corpus of Spontaneous Japanese* (cf. endnote 1 in chapter 6 for CSJ). 75% of BPM in CSJ is H%, 24% is HL%, and 1% is HLH% and LH%. Maekawa (2011), following Kawakami 1963, claims that H% is a "normal rise" and there is a stylistic difference between H% and HL%; H% is found more in formal utterances than HL%. We will come back to the semantic and pragmatic functions of BPM later. Before that, we will discuss the locality of BPM first.

7.1 HOW PROSODIC BOUNDARY IS ALIGNED IN JAPANESE

BPM is important semantically and pragmatically, but due to its positional locality, we need to make it clear what "boundary" is in Japanese. The literature is vague on how to chunk prosodic phrases, as reviewed in chapter 1. Japanese Tones and Break Indices (J_ToBI) by Venditti (1995) was the first proposal to label Japanese prosodic information, based on P&B (1988) as a theoretical background. As discussed in chapter 1, P&B (1988) claim that focus appears at the leftmost position of a prosodic phrase (i.e., the intermediate phrase in their term) and resets a prosodic phrase. In their framework, a boundary is aligned by focal prominence. But Shinya (1999) and Kubozono (2007), among others, empirically refuted P&B's reset theory on the sentence level. Our RPT experiments also show that focal prominence appears in phrase-initial, phrase-mid, and phrase-final positions in spontaneous speech (cf. table 6.7). Also, some phrases do not mark focal prominence at all (cf. figure 6.4 (left)). We consider these as serious problems to the reset theory.

CSJ is labeled by X_JToBI, an extended version of J_ToBI, and phrases are chunked by being measured by range F0. X_JToBI classifies three types of phrases; phrase 1 is word-level, phrase 2 is Accentual Phrase (our φ-phrase), and phrase 3 is Intonation Phrase (our ι-phrase). In J_ToBI, BPM is defined as the phrase-final tone on their Intonation Phrase, but in X_JToBI, it is assigned at the final position of their Accentual Phrase (cf. Koiso (ed.), 2015: 86). The distinction between phrase 2 and phrase 3 derives from whether the pitch range is reset or not (cf. Maekawa, 2011: 40). We have pointed out that it is problematic to chunk phrases by pitch reset. Also, the reason why CSJ chunks boundaries by range F0 is not clear. We wonder whether it is well-motivated to use acoustic cues to chunk boundaries. We argued in chapter 6 that we cannot predict boundaries by acoustic features of maximum F0, duration, or intensity (cf. table 6.4). In our RPT experiments, the correlation between b(oundary)-score and p(rominence)-score is very low (r=0.12 on Person's Correlation). We consider that boundaries and prominences are not so much co-related as believed in traditional literature, and to chunk boundaries by range F0 alone is not the best way to align boundaries, either.

CSJ assumes three types of prosodic boundaries: absolute boundary, strong boundary, and weak boundary (cf. Maekawa, 2011: 135). They classify Japanese conjunctions into three types: absolute, strong, and weak, and prosodic boundaries are classed into three according to the strength of the conjunction they appear with. But it is not clear what motivates this classification, or what these three boundary classes account for. Utterances do not necessarily contain conjunctions, and where there is no conjunction, there is no way to classify types of prosodic boundaries. Maekawa (2011) admits that it is not easy to chunk boundaries in spontaneous Japanese. We find it unconvincing to mark boundaries by types of conjunctions.

In chapter 6, we claim that prosodic boundaries are more or less syntactically controlled (cf. figure 6.2). S and Conjunction are effective boundary cues, and XPs such as NP and VP are also boundary cues in Japanese. There is a cross-linguistic variation in what syntactic categories and non-syntactic categories are possible boundary cues. English, for example, uses higher syntactic categories of S and Conjunction and the non-syntactic category of Discourse Marker as effective boundary cues (cf. figure 7.2), while Japanese also uses lower syntactic categories. English and Japanese are different in how to use non-syntactic categories as a boundary cue: English uses Discourse Markers and Disfluencies effectively to align boundaries, but Japanese uses topic marker more than Discourse Markers (cf. figure 6.2 in chapter 6).

The difference in boundary cues possibly leads to a perception bias cross-linguistically. In our previous study, we found that Japanese learners of English (JEFLs) use lower syntactic cues of XP more than native English learners (cf. Mizuguchi et al., 2017, 2019a). Also, JEFLs are not good at using non-syntactic categories of Discourse Markers and Disfluencies as boundary cues.

7.2 WHAT IS MARKED AS PROMINENT IN JAPANESE

In chapter 1, we have observed that prominence is classed into "alignment/non-focal" prominence and "focal" prominence in the literature. In a left-branching language, the left-most word is labeled strong accent syntactically. Japanese is a left-branching language, and it is expected that prominence

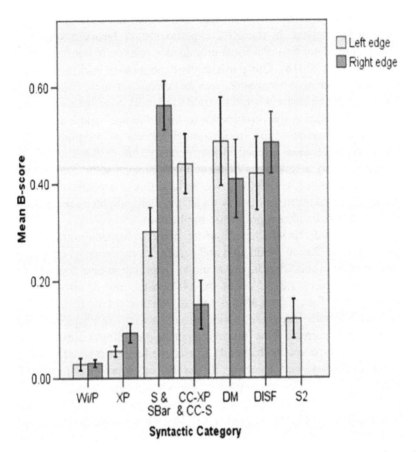

Figure 7.2 Mean B-Score per Category at the Left and Right Edge. *Source*: Cole et al. (2010a: 1161), reprinted with permission from Language Cognitive Processes.

is assigned to the left-most word in a prosodic phrase. Selkirk and Tateishi (S&T) (1988) is a pioneering work on the syntax-prosody interface. Our perception experiments on spontaneous Japanese in chapter 6, however, find that prominences are not restricted to the left-most phrase-initial word and claim for the reconsideration of the idea that prominence is aligned structurally.

Recall that Japanese has many IPs without prominence (cf. table 6.7, figure 6.4). Unlike English, which has Nuclear Stress Rule, Japanese does not demand the utterance-final content word to get stressed. English focus is assigned structurally and is called alignment focus. We assume it is equivalent to "broad focus"(BF) in the sense of intonational literature (cf. Halliday, 1967; Wells, 2006, among others). Japanese probably does not mark BF, and the prominence in Japanese is not assigned syntactically. If we are on the right track in not assuming non-focal prominence aligned syntactically in

Japanese, it is compatible with the fact that there are IPs without prominence marked and perceived. In our RPT experiments on Japanese spontaneous speech, we could not find non-focal prominence marked in our data set, contra Ishihara et al. (2018). Our participants unanimously said that it was no easy matter to perceive prominence in spontaneous Japanese, while English, their L2, was much easier when they tried to identify prominence.

Focal prominence is the prominence to mark "focus," and it is related to the semantic, pragmatic, and informational aspects of prominence. There are many types of focus studied in the literature, but as K&S (2020) point out, there are only a few types of focus that claim prosodic prominence in English. As discussed in chapter 2, K&S claim that it is contrastive (FoC in their term), but not information-new focus, that is highlighted prosodically in English, and FoC invokes a pragmatic implicature.

We have considered whether Japanese marks information-new focus in chapter 3. We conducted production and perception experiments on question-answer pairs and found that Japanese marks information-new focus by local F0 boost (cf. figures 3.2–3.5). But the F0 boost is not so strong that the information-new focus is not always easy to perceive (cf. table 3.2).

We tested whether Japanese marks corrective focus in chapters 4 and 5. Corrective focus corrects some information in the preceding context (cf. (1) for English, (2) for Japanese in chapter 4). Corrective focus is marked by local F0 boost (cf. figure 4.4), and generally, it is perceived as prominent in Japanese (cf. tables 4.2–4.3). There are, however, cases where the focus position does not match the F0 peak (cf. figure 4.3). It is due to the late fall of pitch movement in Japanese and leads to a perception bias among Japanese listeners. We suspect Japanese uses F0-boost and pitch movement to perceive corrective focus, and via an fMRI experiment, we have found that Japanese activates the right dorsolateral prefrontal cortex to process pitch movement while English does not (cf. figure 5.2). We come to the conclusion that Japanese uses local F0 boost as well as pitch movement to mark and process corrective focus.

In chapter 6, we conducted perception experiments on spontaneous Japanese and found that focal prominences are marked by local F0 boost and boundary pitch movement (cf. figures 6.4 (right)–6.6). But here comes a question; why does Japanese use two types of prominence cues? Is there any difference between them? We will consider this question in the next section.

7.3 WHY TWO TYPES OF PROMINENCE CUES IN JAPANESE?

In chapter 2, we introduced Tomioka (2009), who points out that prominence is marked either on subject *KEN* or contrastive particle *WA* in (3).

(3) Q: Dare-ga   ukat-ta-no?
         Who-GA   pass-PAST-Q
         "Who passed?"

     A: KEN-wa/Ken-WA   ukat-ta.
         Ken-TOP   pass-PAST
         " (At least) Ken passed."   (Tomioka, 2009: 119)

The prominence of the subject marks information-new focus, and the prominence of the particle marks contrastive focus in (3A). The former is realized by local F0-boost and the latter is aligned a boundary tone.

Japanese allows prominence alignment not only on content words but also on function morphemes (cf. Oishi, 1959; Kawakami, 1963; Kuroda, 1965; Kuno, 1973; Taniguchi and Maruyama (T&M), 2001; Hara, 2006; Tomioka, 2009, among others). Note that the prominence of a particle does not indicate that the particle alone is in contrast with other particles but the whole phrase to which the particle attaches is in contrast. Tomioka (2009) claims that the effect of the prominence on particles is pragmatic in nature; in (3A), the contrastive reading of *wa* induces an alternative set {Ken passed, Mary passed, John passed. . .}, and implies that other people possibly passed. Without prosodic prominence of topic marker *wa*, (3A) has the thematic reading in the sense of Kuno (1973) (cf. (13) in chapter 2) and there is no implication that other people possibly passed. Following Tomioka (2009), we claim that the prominence of particles induces an alternative set and it provokes pragmatic implicature.

In the framework of alternative semantics, the domain of prominence is determined syntactically (cf. Rooth, 1986, 1992, 2016), and we are going to apply the framework to Japanese. Though the syntactic structure of Japanese is much disputed in the literature and is not yet determined (cf. Kayne, 1994; Fukui, 1995; Saito and Fukui, 1998; Whitman, 2001, among many others), we assume that Japanese is a head-final language. It is often claimed that Japanese lacks D, and Fukui (1995) proposes the category K(ase) as a near-equivalent of D which is the head of Kase Phrase (KP). On this assumption, topic marker *wa*, for example, is the head of the KP in (4a).

(4) a. syntactic labeling: $[_{KP} [_N \text{Ken}]\text{-}[_K \text{wa}]]$
                               TOP

     b. FoC assignment: $[_{KP} [_N \text{Ken-}[_K \text{wa}]]^{\text{FoC}}$

     c. alternative set: $[\![\text{Ken-wa}]\!]^{\text{FoC}} = \{\lambda w. \text{ person y in w} \mid y \in D_{e,t}\}$
                               $= \{\text{Ken, Mary, John} . . .\}$

     d. prosodic prominence: Ken-wa
                               H% /HL%

When the head is prominent, the domain of prominence covers the whole syntactic phrase. If we are on the right track to assume that the domain of prominence FoC is syntactically determined, then the semantic type of an alternative set is subsequently determined; in (4c), for example, the semantic type is the type of KP, that is, $<e,t>$, and the members of the alternative set are contextually determined. Prominence is realized as a pitch movement, and in (4d), H% or HL% is assigned to the head of a topic phrase and marks prominence. Notice here that topic *wa* is not lexically accented and H% / HL% is a boundary tone.

The story is different where content words are highlighted. The number of prominences on content words is much bigger than that on function morphemes (cf. table 6.7 in chapter 6). Pitches of content words are lexically determined as H*+L for accented words and LH for unaccented words (cf. (10) in chapter 2). These lexical pitches are kept when they are highlighted (cf. figures 6.4–6.6). We consider that the domain of a highlighted content word is the word itself. (5) illustrates how *Ken* in (3A) is assigned prominence.

(5)   a. syntactic labeling: $[_{KP} [_N$ Ken$]-[_K$ wa$]]$

                                      TOP

      b. FoC assignment: $[_{KP} [_N$ Ken$]^{FoC} -[_K$ wa$]]$

      c. singleton set: $[\![$Ken^{FoC} -wa$]\!]$ = {λw. person in w | y∈$D_{e,t}$ & |D|=1 }
                                          = {Ken}

      d. prosodic prominence: Ken-wa

                                H*+L with F0-boost

FoC is only assigned to *Ken*. Since the domain of FoC is restricted to the noun, we propose that it induces not an alternative set but the singleton set. Prominence is marked on *Ken* by F0-boost.

We are claiming that both boundary tones and local F0-boost highlight focal prominence, but the focus domains are different. In the literature, Boundary Pitch Movement (BPM) has been considered to carry a function to chunk boundaries (cf. T&M, 2001; Maekawa, 2011, among others), or is simply a stylistic sentence-final contour of younger generations (cf. Kori, 2020: 185). We claim that BPM has a more important role. Note that BPM is not restricted to particles. As T&M themselves point out, Auxiliaries (cf. *teiru* in (6aA)) and Adverbials (cf. *kara* in (6bA)) are also assigned a boundary tone of H%.

(6) a. Q: Suiyo-kara   go-shukuhaku-no   Suzuki-sama   deshou-ka?
         Wednesday-from   POLITE-stay-GEN   Suzuki-Polite   Copula-Q

"(Are you)[1] Mr/Ms Suzuki, staying from Wednesday?"

A: Iie,   Kinyo-kara   tomat-<u>teiru</u>   Suzuki-desu-ga

    no   Friday-from   stay-PROG   Suzuki-Copula-Conj

    "No, (I am) Suzuki, staying from Friday, . . ."

b. Q: Doshite   Tokyo-ni   it-ta-no-desu-ka?

       why   Tokyo-to   go-PAST-COMP-Copula-Q

       "Why did (you) go to Tokyo?"

   A: Gakkai-ga   at-ta-<u>kara</u>   Tokyo-ni

       academic conference-NOM   be-PAST-because   Tokyo-to

       it-ta-no-desu.

       go-PAST-COMP-Copula

       "Because (I) had an academic conference, (I) went to Tokyo."

       (T&M, 2001: 63)

Teiru does not mark a phrase boundary in (6aA). The same is the case with figures 6.5–6.6 in chapter 6. We give the texts and the tones of these figures below.[2]

(7)   Text and tones of figure 6.6

     doshite nihon-jin-wa   supichi-ga   heta-nan-daro-ka

     why   Japanese-person-TOP speech-NOM   poor-COPULA-may-Q

     %L H*+L[+Boost]   L%   %LH[+Boost]L% %LH   L%

     to   kangae-mashi-ta

     COMP   consider-POLITE-PAST

     H%[BPM]   %LH   L%

     "(I) wondered why Japanese are poor at speech."

(8)   Text and tones of figure 6.7

     yuyo   konchu   toshite   atsukau   tema-<u>no</u>

     useful   insect   as   analyze   theme-GEN

     %LH[+BOOST] H*+L[+BOOST] LHL%[BPM] %LH%   %H*+L

                                                     [+BOOST]

                                                     H%[BPM]

     ronbun-o   kai-ta

     paper-ACC   write-PAST

     %L   H*+L   L%

     "(I) wrote an article that analyzes (X) as a useful insect."

In (7), complementizer *to* is perceived as prominent by the listeners of Osaka and NK dialects (see table 6.8 for p-scores). In (8), genitive particle *no* is perceived as prominent by the listeners of Tokyo and Osaka dialects (see table 6.9 for p-scores). Both are aligned boundary tones, but prosodically, these function morphemes do not chunk a clause or a noun phrase. We consider these as counterexamples of the delimitative function of BPM.

7.4 TOWARD FUTURE STUDY

We have argued that local F0 boost and BPM have different functions in Tokyo Japanese. Since our idea is new in the field, there might be counterarguments. We have not accounted for why the listeners of different dialects behave differently in boundary and prominence perception (cf. table 6.7). It may be because pitch movements vary among dialects in Japanese. Nowadays, almost all Japanese can understand Tokyo Japanese, but our RPT study shows that their perception of prosody is not the same. Though we could not discuss tonal scaling in this book, there are dialects with downward scaling (e.g., Tokyo Japanese) and with upward scaling (e.g., NK Japanese, Kumamoto Japanese, Koriyama Japanese, see Igarashi, 2014 for further details). We suspect tonal scaling might affect our perception strategy; we suspect Japanese dialects with upward tone scaling does not use F0 boost to mark prominence, and effective prominence cues vary among dialects. We need to study more dialects on a bigger scale to answer our speculation. We have not considered the difference between BPMs, either. Maekawa (2011) shows that 75% of BPM in CSJ is H% and 24% is HL% and claims the difference between H% and HL% is stylistic. We will leave the study on the variation in BPM for the future.

Cross-linguistic RPT studies are currently in tremendous progress (Bishop et al., 2020 for English, Baumann and Winter, 2018; Baumann and Schumacher, 2020 for German, Cole et al., 2019 for English, French, and Spanish). We hope our RPT study will deepen the discussion of Japanese prosody and lead to a new perspective in the cross-linguistic study of prosody.

NOTES

1. Japanese can delete arguments as far as they are recoverable from the contexts. We put the deleted arguments in the parentheses in the gloss.

2. Our tone assignments are probably different from X_JToBI. *Ronbun* in Figure 6.7 is lexically unaccented and its tones must be LH. But we do not see the tone LH in Figure 6.7. We therefore mark %L in (8).

Appendix A

The following are the stimuli and the contexts of the production experiment in chapter 3.

The experiment consisted of the Exercise session and Set 1 to Set 4.

(N.B. Though the numbers are given in the Arabic numerals below, they are produced in Japanese numerals (cf. Exercise Session for Japanese pronunciation) in the response.)

EXERCISE SESSION

I. Participants practiced reading the following nouns, adjectives, proper nouns, and numbers.

nouns: ume (U, plum), 'uni (A, sea urcin), ane (U, elder sister), 'ani (A, elder brother)
adjectives: amai (U, sweet), a'oi (A, blue)
proper nouns: Berurin (U, Berlin), 'Daburin (A, Dublin)
numbers: 'ichi(one), 'ni(two), 'san(three), 'shi/'yon(four), 'go(five), 'roku(six), 'shichi/'nana(seven), 'hachi(eight), 'kyu(nine), 'zero(zero)
roku roku ni (662), ni ichi ni(212), roku san ni(632), yon go ni(452), roku roku ichi(661), ni ichi ichi(211), roku san ichi(631), yon go ichi (451)

II. Participants practiced giving the answer part (A1, A2, A3) orally, after reading the question part (Q1, Q2, Q3). Questions (4)–(5) are distractors.

II-1. What is it? Type (gives the broad focus context)

    Q1: Nani-ga   suki-desu-ka?
         what-NOM   like-Copula(POL)-Q   "What (do you)[1] like?"
    A1: Aoi   uni-desu.
         blue   sea urchin-Copula(POL) "(I like) blue urchin."

II-2. What Y was it? Type (gives the information new focus context on Modifier)

    Q2: Donna   uni-ga   suki-desu-ka?
         what   sea urchin-NOM   like-Copula(POL)-Q
         "What sea urchin (do you) like?"
    A2: Aoi   uni-desu.
         blue   sea urchin-Copula(POL)   "(I like) blue [+F] urchin."

II-3. X's what was it? Type (gives the information new focus context on Head)

    Q3: Aoi   nani-ga   suki-desu-ka?
         blue what-NOM   like-Copula(POL)-Q   "What blue (do you)
                                                  like?"
    A3: Aoi   uni-desu.
         blue sea urchin-Copula(POL)   "(I like) blue urchin
                                                  [+F]."

II-4. Q4: Yamada-san-no   denwa-bango-wa
           Yamada-Mr/Ms-GEN   telephone number-TOP
           nan-ban-desu-ka?
           what-number-Copula(POL)-Q
           "What is Mr/Ms Yamada's phone number?"
      A4: 022-035-1274 desu.   "(It is) 022-035-1274."

II-5. Q5: Yamada-san-no   denwa-bango-wa 022-035-1274 desu-ka?
           Yamada-Mr/Ms-GEN   telephone-number-TOP Copula(POL)-Q
           "(Is) Mr/Ms Yamada's number 022-035-1274?"

      A6: Iie.   032-035-1274   desu.
           no   Copula(POL) "No. (It is) 032-035-1274."

Set 1

1.	Q1	A:	Amai ume-desu. (UU)
2.	Q1	A:	Amai 'uni-desu. (UA)
3.	Q2 with *ume*	A:	Amai ume-desu. (U[+F]U)
4.	Q2 with *'uni*	A:	Amai 'uni-desu. (U[+F]A)
5.	Q2 with *ume*	A:	A'oi ume-desu. (A[+F]U)
6.	Q2 with *'uni*	A:	A'oi 'uni-desu. (A[+F]A)
7.	Q4	A:	022-025-0274-desu.
8.	Q4	A:	424-724-0424 desu.
9.	Q4	A:	472-772-0274 desu.
10.	Q5 with 022-025-0374	A:	022-025-0274 desu.
11.	Q5 with 424-724-0524	A:	424-724-0424 desu.
12.	Q5 with 427-772-6274	A:	427-772-0274 desu.

Set 2

1.	Q3 with *amai*	A:	Amai ume-desu. (UU[+F])
2.	Q3 with *amai*	A:	Amai 'uni-desu. (UA[+F])
3.	Q3	A:	A'oi 'uni-desu. (AA[+F])
4.	Q3	A:	A'oi ume-desu. (AU[+F])
5.	Q4	A:	022-025-220 desu.
6.	Q4	A:	022-025-779 desu.
7.	Q4	A:	442-742-247 desu.
8.	Q5 with 022-025-2400	A:	022-025-2200 desu
9.	Q5 with 022-025-0799	A:	022-025-7799 desu
10.	Q5 with 422-742-7474	A:	422-742-2474 desu
11.	Dare-ga ki-masu-ka? "Who will come?"	A:	'Daburin-no ane-desu. (AU) "(My) sister in Dublin (will)."
12.	same as #11	A:	'Daburin-no 'ani-desu. (AA)

Set 3

1.	Q4	A:	022-025-7700 desu.
2.	Q4	A:	022-025-2299 desu.
3.	Q4	A:	022-025-7722 desu.
4.	Q5 with 022-025-7100	A:	022-025-7700 desu.
5.	Q5 with 022-025-2599	A:	022-025-2299 desu.
6.	Q5 with 022-025-1722	A:	022-025-7700 desu.
7.	Q1	A:	A'oi ume desu. (AU)
8.	Q1	A:	A'oi 'uni desu. (AA)
9.	Doko-no ane-ga ki-masu-ka? "Which sister came?"(contrastive)	A:	Berurin-no ane-desu. "Berlin's sister came." (U[+F]U)

10. same as #9   A: 'Daburin-no ane-desu.(A[+F]U)
11. Doko-no ani-ga ki-masu-ka?   A: Berurin-no 'ani-desu.
     "Which brother   "Berlin's brother came." (U[+F]
     came?"(contrastive)   A)
12. same as #11   A: 'Daburin-no 'ani-desu. (A[+F]A)

Set 4

1.   Q5 with 022-025-2577   A: 022-025-2277 desu.
2.   Q5 with 024-224-7424   A: 022-224-0424 desu.
3.   Q5 with 224-424-7924   A: 022-424-7224 desu.
4.   same as #11 in Set 2   A: Berurin-no ane-desu. (UU)
5.   same as #11 in Set 2   A: Berurin-no 'ani-desu. (UA)
6.   'Daburin-no dare-ga   A: 'Daburin-no 'ani-desu.
     suki-desu-ka?   (AA[+F])
     "Who in Dublin (do you) like?"
7.   same as #6   A: 'Daburin-no ane-desu.
                                             (AU[F+])
8.   Berurin-no dare-ga   A: Berurin-no 'ani-desu.
     suki-desu-ka?   (UA[+F])
     "Who in Dublin (do you) like?"
9.   same as #8   A: Berurin-no ane-desu. (UU[F+])
10. Q4   A: 022-025-2277 desu.
11. Q4   A: 022-224-2277 desu.
12. Q4   A: 022-424-7224 desu.

NOTE

1. Japanese can delete arguments as far as they are recoverable from the context.
We put *I/you* for the subject in the gloss tentatively.

Appendix B

The following shows the materials for Perception Experiments in chapter 3. The target materials consist of two words or numbers (marked as W1 and W2 below). The task is to listen to the recorded material once, and mark which of [W1] or [W2] is highlighted prosodically, or mark [B] in case participants perceive no prosodic highlight. There are 96 materials to judge: 48 contrastive foci (CF), 24 broad foci (BF), and 24 distractors, that is, numbers. Most of the materials were recorded in the production experiment in Appendix A by male native speakers of Tokyo Japanese, and the rest of the materials were recorded by two male native speakers of Tokyo Japanese. The experiment is completed in about 30 minutes. Participants took rest in the mid of the experiment, when necessary.

EXERCISE SESSION

Listen to the recording once and judge which of W1 and W2 is highlighted. When you hear no prosodic highlight, mark [B].

(1). 022-025-2(W1) 2(W2) desu.
(2). Daburin-no (W1) ani (W2) desu.
(3). Amai (W1) uni (W2) desu.

MATERIALS

The numbers are written in the Arabic numerals in the following material list, but they are produced in Japanese (refer to Exercise session in Appendix A for Japanese pronunciation) in the experiment.

Since we cannot attach sound files here, we provide lexical accent and focus information in the square bracket; [AA], for instance, shows both W1 and W2 are Accented words, and [U[+F]U] marks W1 as Unaccented word with prosodic highlight. [Dis] is a distractor.

1. [AA]   Daburin-no(W1) ani(W2) desu.
2. [A[+F]A]   Aoi(W1) uni(W2) desu.
3. [Dis]   022-025-2(W1)2(W2)00 desu.
4. [Dis]   022-025-2(W1)2(W2)99 desu.
5. [Dis]   022-025-7(W1)7(W2)99 desu.
6. [UA[+F]]   Amai(W1) uni (W2) desu.
7. [Dis]   022-025-2(W1)2(W2)77 desu.
8. [U[+F]A]   Amai(W1) uni(W2) desu.
9. [U[+F]A]   Amai(W1) uni(W2) desu.
10. [AA[+F]]   Aoi(W1) uni(W2) desu.
11. [Dis]   022-025-2(W1)2(W2)00 desu.
12. [Dis]   224-424-7(W1)2(W2)24 desu.
13. [Dis]   022-025-7(W1)7(W2)22 desu.
14. [AA[+F]]   Aoi(W1) uni(W2) desu.
15. [Dis]   024-224-0(W1)4(W2)24 desu.
16. [Dis]   022-025-7(W1)7(W2)99 desu.
17. [A[+F]A]   Aoymana-no(W1) ani(W2) desu.
18. [Dis]   422-742-2(W1)4(W2)74 desu.
19. [Dis]   427-772-0(W1)2(W2)74 desu.
20. [U[+F]U]   Oomiya-no(W1) ane(W2) desu.
21. [UA]   Amai (W1) uni(W2) desu.
22. [Dis]   472-772-0(W1)2(W2)74 desu.
23. [Dis]   022-025-0(W1)2(W2)74 desu.
24. [AA[+F]]   Aoyama-no(W1) ani(W2) desu.
25. [Dis]   022-025-(W1)7(W2)99 desu.
26. [Dis]   022-025-2(W1)2(W2)99 desu.
27. [AU]   Daburin-no(W1) ane(W2) desu.
28. [AU[+F]]   Aoymana-no(W1) ane(W2) desu.
29. [Dis]   224-424-7(W1)2(W2)24 desu.
30. [Dis]   022-025-7(W1)7(W2)99 desu.
31. [Dis]   424-724-0(W1)4(W2)24 desu.
32. [Dis]   442-742-2(W1)4(W2)74 desu.
33. [UU[+F]]   Amai(W1) ume(W2) desu.
34. [Dis]   472-772-0(W1)2(W2)74 desu.
35. [AU[+F]]   Aoi(W1) ume(W2) desu.
36. [Dis]   442-742-2(W1)4(W2)74 desu.
37. [AU[+F]]   Aoi(W1) ume(W2) desu.

38. [UU]   Amai(W1) ume(W2) desu.
39. [UU[+F]]   Amai(W1) ume(W2) desu.
40. [U[+F]U]   Amai(W1) ume(W2) desu.
41. [Dis]   022-025-7(W1)7(W2)00 desu.
42. [UU]   Amai(W1) ume(W2) desu.
43. [Dis]   022-025-0(W1)2(W2)74 desu.
44. [Dis]   424-724-0(W1)4(W2)24 desu.
45. [Dis]   022-025-7(W1)7(W2)00 desu.
46. [UA[+F]]   Amai(W1) uni(W2) desu.
47. [Dis]   022-025-0(W1)2(W2)74 desu.
48. [UU]   Berurin-no(W1) ane(W2) desu.
49. [Dis]   024-424-0(W1)4(W2)24 desu.
50. [UU[+F]]   Oomiya-no(W1) ane(W2) desu.
51. [AU]   Aoyama-no(W1) ane(W2) desu.
52. [Dis]   022-025-2(W1)2(W2)77 desu.
53. [Dis]   022-025-2(W1)2(W2)77 desu.
54. [Dis]   424-724-0(W1)4(W2)24 desu.
55. [UA]   Oomiya-no(W1) ani(W2) desu.
56. [U[+F]U]   Amai(W1) ume(W2) desu.
57. [Dis]   022-025-2(W1)2(W2)99 desu.
58. [Dis]   442-742-2(W1)4(W2)74 desu.
59. [Dis]   024-224-0(W1)4(W2)24 desu.
60. [Dis]   024-224-0(W1)4(W2)24 desu.
61. [UU[+F]]   Berurin-no(W1) ane(W2) desu.
62. [Dis]   424-724-0(W1)4(W2)24 desu.
63. [Dis]   224-424-7(W1)2(W2)24 desu.
64. [AA[+F]]   Daburin-no(W1) ani(W2) desu.
65. [UU]   Oomiya-no(W1) ane(W2) desu.
66. [UA]   Berurin-no (W1) ani(W2) desu.
67. [Dis]   022-025-0(W1)2(W2)74 desu.
68. [Dis]   022-025-2(W1)2(W2)00 desu.
69. [Dis]   427-772-0(W1)2(W2)74 desu.
70. [A[+F]U]   Aoi(W1) ume(W2) desu.
71. [AU[+F]]   Daburin-no(W1) ane(W2) desu.
72. [Dis]   022-025-2(W1)2(W2)00 desu.
73. [AA]   Aoi (W1) uni(W2) desu.
74. [AA]   Aoi (W1) uni(W2) desu.
75. [Dis]   022-025-7(W1)7(W2)00 desu.
76. [A[+F]U]   Aoyama-no (W1) ane(W2) desu.
77. [U[+F]A]   Oomiya-no (W1) ani(W2) desu.
78. [UA]   Amai(W1) uni(W2) desu.
79. [Dis]   022-025-2(W1)2(W2)77 desu.

80. [A[+F]A]   Aoi(W1) uni(W2) desu..
81. [A[+F]U]   Daburin-no(W1) ane(W2) desu.
82. [UA[+F]]   Berurin-no(W1) ani(W2) desu.
83. [U[+F]A]   Berurin-no(W1) ani(W2) desu.
84. [A[+F]A]   Daburin-no(W1) ani(W2) desu.
85. [A[+F]U]   Aoi (W1) ume(W2) desu.
86. [Dis]   22-025-7(W1)7(W2)00 desu.
87. [Dis]   024-224-0(W1)4(W2)24 desu.
88. [U[+F]U]   Berurin-no(W1) ane(W2) desu.
89. [U[+F]A]   Oomiya-no(W1) ani(W2) desu.
90. [Dis]   022-025-2(W1)2(W2)99 desu.
91. [Dis]   022-025-2(W1)2(W2)77 desu.
92. [AA]   Aoyama-no(W1) ani(W2) desu.
93. [Dis]   022-025-7(W1)7(W2)22 desu.
94. [Dis]   022-025-7(W1)7(W2)22 desu.
95. [AU]   Aoi (W1) ume(W2) desu.
96. [Dis]   022-025-7(W1)7(W2)22 desu.

Appendix C

One male speaker of Midwest American English and one male speaker of Tokyo Japanese recorded a series of 10 question-answer pairs which contained a 10-digit number of the form XXX-XXX-XXXX (cf. (1)). In each of the responses, one of the 10 numbers was produced with corrective focus.

(1) a. Japanese material

    Q: Yamada-san-no   denwa-bango-wa   215-413-5623 desu-ka?

       Yamada-Mr/Ms-GEN   phone-number-TOP   Copula(POL)-Q

       "Is Yamada's number 215-413-5623?"

    A: Iie.   215-413-6623 desu.

       no.   Copula(POL)

       "No. (It is) 215-413-6623."

  b. English material

    Q: Is Mary's number 215-413-5623?

    A: No, the number is 215-413-6623.

In the experiment, after the exercise session, subjects listened to each recording and marked the word that they heard as corrective on a transcript. They listened to randomized 30 phone-number strings, that is, each contrastive digit three times. The experiment was completed in about 30 minutes.

The following are the materials with corrective focus, and they are numbered CF1, CF2, . . ., and CF10. The number shows the location of correction in the 10-digit number sequence. In Japanese, the contrastive numbers are followed by *desu* (Copula), and in English, they are preceded by "The number is," to make the answer grammatical (cf. (1)).

CF1	3[+F]15-418-5623
CF2	22[+F]5-418-5623
CF3	216[+F]-418-5623
CF4	215-5[+F]18-5623
CF5	215-42[+F]8-5623
CF6	215-419[+F]-5623
CF7	215-418-6[+F]623
CF8	215-418-57[+F]23
CF9	215-418-563[+F]3
CF10	215-418-5628[+F]

Appendix D[1]

We used 80 English and Japanese broad focus and corrective focus 10-digit numbers (20X2 (broad/contrastive) X2 (Japanese/English)), recorded by one male native speaker of Midwest American English and one male native speaker of Tokyo Japanese. Before the experiment, participants received instruction about the distinction between broad and corrective focus in Question-Answer dialogues, like (1) and (2).

(1) English materials

    a.   CF condition

        Q: Is Mary's number 215-418-5623?

        A: No, the number is 215-417-5623.

    b.   BF condition

        Q: What is Mary's number?

        A: The number is 215-417-5623.

(2) Japanese material

    a.   CF condition

        Q: Yamada-san-no   denwa-bango-wa   215-413-5623

        Yamada-Mr/Ms-GEN   phone-number-TOP

        desu-ka?

        Copula (POL)-Q

        "Is Yamada's number 215-413-5623?"

---

[1] Table in appendix D has no caption and is purposefully absent from the list of tables and figures.

A: Iie.   215-413-6623 desu.

no   Copula(POL)

"No. (It) is 215-413-6623."

b.   BF/Whole sentence condition

Q: Yamada-san-no   denwa-bango-wa   nan-ban

Yamada-Mr/Ms-GEN   phone-number-TOP   what-number

desu-ka?

Copula(POL)-Q

"What is Yamada's number?"

A: 215-413-6623 desu.

Copula(POL)

"(It) is 215-423-6623."

Participants hear only the answer part, and judge whether or not the audio stimuli contained corrective focus. The "Yes" and "No" responses were provided by pressing the button with the index and the middle fingers of the left hand. They sat for a practice session before the experiment. The experiment was completed in about an hour and a half including the experiment explanation, consent credit, and post-experiment questions.

The following are the materials used in the experiment, with the onset and interval time settings.

Japanese material	#	onset (sec.)	Interval (sec.)	English material	#	onset (sec.)	interval (sec.)
215 418 5423	cf-1	0		216 418 5623	bf-e1	0	
215 412 5623	cf-2	10	10	215 418 5633	cf-e1	18	18
615 418 5623	bf-1	20	10	215 428 5623	bf-e2	42	14
215 438 5623	bf-2	30	10	216 418 5623	cf-e2	47	5
213 418 5623	cf-3	44	14	315 418 5623	bf-e3	62	15
215 418 5423	bf-3	54	10	215 418 5633	bf-e4	72	10
215 418 5629	cf-4	64	10	215 418 6623	cf-e3	77	5
245 418 5623	bf-4	78	14	315 418 5623	bf-e5	92	15
215 418 5923	cf-5	96	18	215 219 5623	cf-e4	97	5
615 418 5623	bf-5	106	10	215 518 5623	bf-e6	122	15
215 417 5623	cf-6	116	10	216 418 5623	bf-e7	132	10
215 218 5628	bf-6	126	10	215 418 5633	cf-e5	137	5
212 418 5623	bf-7	136	10	315 418 5623	bf-e8	166	29
215 418 5624	cf-7	146	10	315 418 5623	cf-e6	171	5
245 418 5623	bf-8	160	14	216 418 5623	cf-e7	176	5
295 418 5623	cf-8	174	14	225 418 5623	bf-e9	186	10
212 418 5623	bf-9	184	10	215 219 5623	cf-e8	191	5
217 418 5623	cf-9	194	10	215 428 5623	bf-e10	206	15

(continued)

Japanese material	#	onset (sec.)	Interval (sec.)	English material	#	onset (sec.)	interval (sec.)
215 412 5623	bf-10	208	14	215 418 5723 cf-e9		211	5
315 418 5623	cf-10	218	10	215 528 5623 cf-e10		226	15
215 218 5628	bf-11	232	14	215 419 5623 bf-e11		236	10
215 518 5623	cf-11	242	10	215 418 5723 bf-e12		260	14
215 438 5623	bf-12	252	10	225 418 5623 cf-e11		265	5
915 418 5623	cf-12	270	18	215 418 6623 bf-e13		296	29
215 718 5623	bf-13	284	14	215 518 5623 cf-e12		301	5
215 418 1623	cf-13	298	14	215 418 5723 cf-e13		306	5
215 418 5423	bf-14	308	10	215 419 5623 bf-e14		320	14
245 418 5623	cf-14	318	10	215 418 5723 bf-e15		334	14
215 438 5623	cf-15	328	10	315 418 5623 cf-e14		344	10
215 418 9623	cf-16	338	10	225 418 5623 bf-e16		354	10
215 418 1623	bf-15	348	10	215 518 5623 cf-e15		372	18
215 458 5623	cf-17	362	14	215 528 5623 cf-e16		382	10
215 418 1623	bf-16	372	10	215 418 5633 bf-e17		387	5
215 718 5623	bf-17	386	14	215 418 6623 cf-e17		402	15
215 318 5623	cf-18	404	18	215 418 5624 bf-e18		407	5
245 418 5623	bf-18	430	16	215 418 5624 cf-e18		426	19
215 418 5653	cf-19	440	10	215 518 5623 bf-e19		431	15
215 418 5653	bf-19	450	10	225 418 5623 cf-e19		446	15
215 418 5653	bf-20	460	10	215 418 6623 bf-e20		451	5
215 418 5643	cf-20	470	10	215 418 5624 cf-e20		466	15

Appendix E

The following are the transcripts of the RPT experiment materials in chapter 6, extracted from *Corpus on Spontaneous Japanese* (CSJ), released by the National Institute for Japanese Language and Linguistics (NINJAL). In our experiments, the text is written in Japanese *Hiragana*, that is, Japanese cursive characters, but we transcribe the texts in the Roman letters below, segmented on the morpheme level. Punctuations and capital letters are not marked in the transcripts. Japanese allows argument deletion on the context recovery basis, so we insert arguments in the parentheses to make our translation of the experiment materials more English-like.

ABBREVIATIONS

SUBJ: Subjective, ACC: Accusative, LOC: Locative, GEN: Genitive, COMP: Complementizer, PAS: Passive, PROG: Progressive, ASP: Aspect, NEG: Negation, CL: Classifier, PL: Plural, POL: Polite, HON: Honorific, HUM: Humble, TOP: Topic Marker, Q: Question Marker, PART: Particle, DM: Discourse Marker, DISF: Disfluency

EXERCISE (FILE # S00F0031)

mazu	watashi	no	sukina	mono	inu	desu	kyonen	no
first	I	GEN	like	thing	dog	Copula	last year	GEN

ima	goro	made	shettolando shipudoggu	toiu	inu	o
now	time	till	Shetland sheepdog	so-called	dog	ACC

kat	tei	mashi	ta	mukashi	hayat	ta
keep	Prog	Copula(POL)	Past	old days	popular	Past

kori	no	kogata	ban	mitaina	inu	no	koto
Collie	GEN	small size	size	like	dog	GEN	thing

desu	watashi	no	chugakko	nyugaku	to
Copular(POL)	I	GEN	middle school	entrance	with

dojini	kat	temorat	ta	n	desu	ga
at the same time	buy	receive	PAST	COMP	Copula(POL)	but

"First, my favorite (is) a dog. I kept a dog called Shetland sheepdog till around this time last year. (It is) a small-sized Collie, which was once popular. (They) bought (the dog) for me when I entered middle school. . . ."

#1 (File #A01f0055)

e	chocho	tancho	no	chigai	wa	e
eh	major key	minor key	GEN	difference	TOP	eh

seijin	no	ba'ai	wa	hikakuteki	yoini	ano	hudan	kara
adult	GEN	case	TOP	relatively	easily	DM	usually	from

kiki	wake	tei	rare	to	iwa	re	tei	masu
hear	distinguish	PROG	PAS	COMP	say	PAS	PROG	Copula(POL)

mata	kaigai	no	kenkyu	dewa	shi	go	sai	ji	yoji
and	overseas	GEN	research	in	four	five	year	child	small child

wa	ryosha	no	benbetsu	ga	kano	de	aru	to
TOP	both	GEN	distinction	SUBJ	possible	Copula	be	COMP

iu	hokoku	mo	ari	masu		dewa	itsu	goro	kara	kano
say	report	too	be	Copula(POL)		if so	when	time	from	possible

na	no	desho		ka	aruiwa	ningen	no	hattatsu
Copula	COMP	Copula(POL)		Q	or	human	GEN	development

no	shoki	no	dankai	kara	mi	rareru	yona	e
GEN	early days	GEN	stage	from	see	PAS	like	DM

keiko	na	no	desho	ka ...
tendency	Copula	COMP	Copula(POL)	Q

"Eh, as for the distinction between major keys and minor keys, it is said that adults can usually perceive it relatively easily. Also, in the research overseas, there is a report that small children at the age of four or five years can distinguish between both keys. Then from when is it possible? Or (is the distinction), eh, the tendency observed from the early stage of human development?"

#2 (File #A01f0090)

atarashi'i	go	o	tsukuri	dasu	katei	o
new	word	ACC	produce	get out	process	Acc

hukugo	to	yobi	masu	kono	katei	ni	oite
compound	COMP	call	Copula(POL)	this	process	LOC	in

tango	no	kumiawase	ni	yotte	wa	a'o	to	sora
word	GEN	combination	LOC	by	TOP	blue	and	sky

de	aozora	no	yoni	ni	ban	me	no
by	blue sky	GEN	like	two	order	SUFFIX	GEN

go	to	onsetsu	o	daku	on	ka	suru
word	initial	syllable	ACC	voiced	sound	SUFFIX	do

gensho	rendaku	ga	mi	rare	masu	mata ...
phenomenon	euphonic change	SUBJ	see	PAS	Copula(POL)	also

"(We) call the process to generate new words compound. In this process, depending on the combination of words, as in *ao-zora* consisting of *ao* and *sora*, the euphonic change of an unvoiced sound to a voiced sound is observed. Also. . . ."

#3 (File #A01f0122)

sono	kekka	kankoku	chugoku	tai	eigo	furansugo
that	result	Korea	China	Thai	English	French

washa	ni	tuite	kihon	teki	ni	wa	nagasa	de
speaker	LOC	about	basics	SUFFIX	LOC	TOP	length	by

handansa	reru	cho	tan	bo'in	no	dotei
judge	PAS	long	short	vowel	GEN	identification

ni	oite	akusento	gata	no	eikyo	ga	mi
LOC	in	accent	type	GEN	influence	SUBJ	see

rare	oto	no	taka	sa	tsuyo	sa	no
PAS	sound	GEN	high	SUFFIX	strong	SUFFIX	GEN

eikyo	ga	shisasa	re	mashi	ta	gutaiteki
influence	SUB	suggest	PAS	Copula(POL)	PAST	specifically

ni	wa	takai	tsuyoi	bo'in	o	yori	nagaku	kiku
LOC	TOP	high	strong	vowel	ACC	much	long	hear

keiko	ga	mi	rare	mashi	ta.
tendency	SUBJ	see	PAS	Copula(POL)	Past

"As a result, in the identification of long and short vowels judged basically by length, the influence of accent types is observed among the speakers of Korean, Chinese, Thai, English, and French, and the height and the intensity of vowels were pointed out to influence (the perception). Concretely, (it was) observed that these listeners tend to perceive higher and stronger vowels as longer."

#4 (File #A01m0022)

etto	ma	wareware	no	mokuteki	to	shi mashi	te
DM	DM	our	GEN	object	as	do Copula(POL)	ASP

wa	genzai	aki	mah	go	zonji	no	yoni	keitai
Top	present	DISF	DISF	SUFFIX	know(HUM)	GEN	as	mobile

denwa	no	keiyaku	sha	su	mo	zoka	shi	teki
telephone	GEN	contract	person	number	also	increase	do	PROG

mashi	te	eh	koritsu	teki	na	onsei	asshuku	no
Copula(POL)	and	DM	effective	SUFFIX	Copula	sound	compression	GEN

gijutsu	ga	hitsuyo	to	nat	teki	tei	masu	mata
skill	SUBJ	necessary	COMP	become	PROG	PROG	Copula(POL)	and

hinshitsu	ni	taisuru	human	mo	o'oku	a	ko
quality	LOC	with	dissatisfaction	also	many	DISF	high

hinshitsu	ka	ni	taisuru	yokyu	mo	takamat	tei
quality	SUFFIX	LOC	for	demand	also	increase	PROG

masu	mah	saikin	wa	pasonaru	konpyuta	nado	ni
Copula(POL)	DM	recent	TOP	personal	computer	and so on	LOC

mochi'i	rareru …
use	PAS

"Eh, the year, as for our object, now, as you know, the number of contractors of mobile phones is increasing, and eh, the skill of the effective sound compression is getting on demand. Also, dissatisfaction with the quality is many in number, ah, the demand for high quality is mounting. Eh, recently, (the stuff)used for personal computer and the like. . . ."

#5 (File #A01m0037)

soreto	ma	inritsu	ni	desu		ne	sono
and	DM	prosody	LOC	Copula(POL)		PART	DM

washa	no	kojin	sei	no	han'ei	to	iu
speaker	GEN	individual	SUFFIX	GEN	reflection	COMP	say

mono	ga	ima	made	domo	husoku	shi	tei	nakat	ta
thing	SUBJ	now	by	DM	lack	do	PROG	NEG	PAST

ka	na	to	iu	koto	de	picchiba	ni	arawareru
Q	Part	COMP	say	COMP	Copula	pitch bar	LOC	realize

kojin	sei	no	koto	o	eh	chotto	kyo
individual	SUFFIX	GEN	COMP	Acc	DM	a little	today

go	hokoku	shi	tai	to	omoi	masu
SUFFIX(HON)	report	do	want	COMP	think	Copula(POL)

eh	chotto	iregula	desu	kedomo	saki	ni
DM	a little	irregular	Copula(POL)	but	before	LOC

demonsutoreshon	o	ik	ko	ki'i	teitadaki	tai	to
demonstration	ACC	one	CL	listen	get(HON)	want	COMP

omoi	masu
think	Copula(POL)

"And, ah, as for the prosody, the year, considering that the realization of the speaker's personality has been, well, scarce, (I) would like to report today, eh, the personality realized in the pitch bar. Though (it is) a little bit irregular, I would like you to listen to one demonstration first."

#6 (File #A06f0028)

etto	eh	mazu	hajimeni	chotto	teisei	nan	desu
DM	DM	first	first	a little	correction	Copula	Copula(POL)

ga	etto	yokoshu	no	etto	91	peji	no	6	no	kosatsu
but	DM	proceedings	GEN	DM	91	page	no	6	no	discussion

ga	5	no	kosatsu	desu		tsuzuki		mashi
SUBJ	5	GEN	discussion	Copula(POL)		continuance		Copula(POL)

te	92	peji	no	7	no	matome	to	iu	no
Conj	92	page	GEN	7	GEN	summary	COMP	say	COMP

ga	6	desu		bango	ga	hito	tsu	zutsu	zurete	shimai
SUBJ	6	Copula(POL)		number	SUBJ	one	CL	each	slide	end(POL)

mashi		ta	ato	mo	it	ten	sanko		bunken		etto
Copula(POL)		Past	and	also	one	CL	reference		references		DM

okada	eh		1995	ga		onodera	1995	no	machigai	desu
okada	DM		1995	SUBJ		onodera	1995	GEN	error	Copula(POL)

teisei		o	yoroshiku		onegai	itashi		masu
correction		ACC	please		please	do(HUM)		Copula(POL)

dewa	hajime	sase			teitadaki		masu
then	begin	CAUSATIVE			get(HUM)		Copula(POL)

"Well, eh, first, (I'd like to) make minor corrections, well, eh, discussion 6 on page 91 of the proceedings is discussion 5. Next summary 7 is 6. The numbers are shifted one by one. One more (typo) is, eh, Okada 1995, eh, in the references is Onodera 1995. Please make corrections. Well, let me start."

#7 (File #S00f0014)

ano	watashi	wa	dochira	ka	to		iu	to
well	I	TOP	which	Q	COMP		say	COMP

bijutsukan	toka		myujikaru		toka	souiu	mono
museum	and so on		musical		and so on	such	thing

ga	tottemo	ma	ima	demo	suki	na	n
SUBJ	much	DM	now	even	like	Copula	COMP

desu		ga	shujin	wa	souiu	mono	ga	wakara
Copula(POL)		but	master	TOP	such	thing	SUBJ	understand

nai	n	desu	ne	de	souiu	koto	mo
NEG	COMP	Copula(POL)	PART	and	such	thing	also

at	te	eh	rizoto	ha	dakara	ryoko	ni
be	and	DM	resort	SUFFIX	because	trip	LOC

iku	to	moh	goruhu	bakku	o	katsui	de	anoh
go	when	DM	golf	bag	ACC	carry	and	DM

hikoki	ni	noru	tte	iu	yona	anoh	souiu
airplane	LOC	get on	COMP	say	like	DM	such

ryoko	o	koko	su	nen	wa	suru	yoni	nat
trip	ACC	here	some	year	TOP	do	like	become

te	ki	mashi	ta
COMP	come	Copula(POL)	PAST

"Well, if anything, I liked, actually I do like now, museums and musicals and the like, very much, but my husband cannot understand those things. And because of that, well, (he is a) resort party. So, when he goes on a journey, he carries, well, his golf bag and gets on a plane. (We have been) going on such trips, ah, these years."

#8 (File #S00f0082)

ehtoh	sotsugyou	ronbun	kai	ta	n	desu
DM	graduation	thesis	write	PAST	COMP	Copula(POL)

kedomo	daimei	wa	kaigara	mushi	oyobi	abura	mushi	ni
but	title	TOP	shell	insect	and	oil	I	nsect LOC

okeru	senshoku	noryoku	ni	tsuite	de	anoh
of	chromosomal	ability	LOC	on	Copula	DM

yoku	kaigara	mushi	ya	abura	mushi	tte	iu	no
often	shell	insect	and	oil	insect	and so on	say	thing

wa	gai	chu	toshite	anoh	yononaka	no	hito
TOP	harm	insect	as	DM	society	GEN	person

ni	wa	ninchi	sare	teiru	n	desu	kedo
LOC	TOP	recognize	PASS	PROG	COMP	Copula(POL)	but

watashi	wa	sono	gai	chu	o	yuyo	konchu
I	TOP	DM	harm	insect	ACC	useful	insect

to	shite	atsukau	tema	no	ronbun	o	kai	te …
COMP	as	handle	theme	GEN	article	ACC	write	and

"Well, I wrote a graduation thesis, and the title is 'On the chromosomal ability of scale insects and aphids.' Well, scale insects and aphids are often recognized as harmful insects by, you know, people in society, but I wrote an article whose theme is to argue that those harmful insects are useful. . . ."

#9 (File #S00f0095)

watashi	wa	tsukemono	ga	tottemo	dai	suki	de
I	TOP	pickles	SUBJ	very	much	like	and

ehto	nakademo	takuan	ga	ichi	ban	suki
DM	above all	pickled radish	SUBJ	one	number	like

na	n	desu	kedo	mo	sore	wa	yahari
Copula	COMP	Copula(POL)	but	and	that	TOP	after all

chi'isai	toki	kara	anoo	i'e	de	takuan	o
small	time	from	DM	house	LOC	pickled radish	ACC

haha	ga	tuke	tei	ta	sei	mo	at	te	anoh
mother	SUBJ	pickle	PROG	PAST	because	also	be	and	DM

hijo	ni	ichi	nichi	ni	tai	ryo
extreme	SUFFIX	one	day	LOC	much	quantity

no	takuan	o	taberu	n	desu	kedomo
GEN	pickled radish	ACC	eat	COMP	Copula(POL)	but

"I like pickles very much, and eh, I like pickled radish most of all. That is because my mother had pickled radish at home since I was small. Ah, (I) eat a large amount of pickled radish every day."

#10 (File #S00f0208)

gai	koku	jin	no	supichi	wa	amerika	jin	ya
foreign	country	people	GEN	speech	TOP	America	people	and

igirisu	jin	no	supichi	wa	hakkiri	wakaru	noni
England	people	GEN	speech	TOP	clearly	understand	but

nihon	jin	no	supichi	wa	hotondo	subete	no
Japan	people	GEN	speech	TOP	almost	all	GEN

hito	no	nihon	jin	no	supichi	ga	nihon	go
people	GEN	Japan	people	GEN	speech	SUBJ	Japan	language

o	hanashi	teiru	noni	yoku	wakara	nai	kore
ACC	speak	PROG	but	well	understand	NEG	this

wa	doshite	daro	doshite	nihon	jin	wa	supichi
TOP	why	may	why	Japan	people	TOP	speech

ga	heta	nan	daro	ka	to	kangae	mashi	ta.
SUBJ	poor	Copula	may	Q	COMP	consider	Copula(POL)	PAST

"(I can) understand the speech by foreign people, such as Americans and Englishmen, clearly, but as for the speech by Japanese, (I) cannot understand speech by almost all of the Japanese people, though they speak Japanese. Why does this happen? (I) considered why Japanese are poor at speeches."

#11 (File #S02m0101)

ma	sono	ko	o	shit	ta	tte	iu	no	wa
DM	that	child	ACC	know	PAST	COMP	say	COMP	TOP

ani	ga	chodo	on'naji	juku	ni	huta	tsu
elder brother	SUBJ	just	same	cram school	LOC	two	CL

ue	de	i	mashi	te	sono	ko	no	o
elder	Copula	be	Copula(POL)	and	that	child	GEN	PREFIX

ne	san	mo	ani	to	on'naji	gakunen
elder sister	SUFFIX(POL)	also	elder brother	with	same	grade

de	on'naji	juku	ni	i	ta	n	desu
Copula	same	cram school	LOC	be	PAST	COMP	Copula(POL)

ne	de	tamatama	sono	juku	tte	no	ga
PART	and	by chance	that	cram school	COMP	thing	SUBJ

huta	tsu	hanareta	kyoudai	tte	iu	no	ga
two	CL	away	brother	COMP	say	thing	SUBJ

hijoni	ooi	juku	de	ani	no	sedai	no
much	many	cram school	Copula	elder brother	GEN	generation	GEN

ototo	imoto	ga	boku	no	sedai	ni
younger brother	younger sister	SUBJ	I	GEN	generation	LOC

chodo	iru	tte	iu	hanashi	de	kyoudai	banashi	de
just	be	COMP	say	story	Copula	brother	story	with

moriagaru	tte	iu	koto	ga	sono	jibun	tachi	no
liven up	COMP	say	COMP	SUBJ	DM	self	PL	GEN

kyoudai	kan	de	yoku	ari	mashi	ta.
brother	among	with	often	happen	Copula(POL)	Past

"Well, (the reason why) I came to know the girl is, my elder brother happened to study at the same cram school by coincidence, and the sister of that girl was also in the same cram school in the grade of my brother. And the cram school was the school where there were many brothers and sisters two years older than I was. The younger brothers and sisters of my elder brother's grades happened to be in my grade. Well then, we often had a lively talk of our brothers and sisters."

#12 (File #S03m0141

ehtto	desu		ne	ehto	uchi		no	sugu
DM	Copula(POL)		PART	DM	my house		GEN	very

chikaku	ni	wa	desu		ne	ushi	goya	ga
neighborhood	LOC	TOP	Copula(POL)		PART	cow	stable	SUBJ

ari masu		de	eh	kono	apato	o	erabu	toki	ni
be Copula(POL)		and	DM	this	apartment	ACC	choose	time	LOC

kettei	innshi	no	hito	tsu	ga	kono	ushi	goya	deshi
decision	factor	GEN	one	CL	SUBJ	This	cow	stable	Copula(POL)

ta	de	naze	ka	to	iu	to	eh	sono	ushi	goya
PAST	and	why	Q	COMP	say	COMP	DM	that	cow	stable

no	mae	o	to'ot	ta	toki	ni	desu		ne
GEN	front	ACC	pass by	PAST	time	LOC	Copula(POL)		PART

eh	asa	warito	hayai	jikan	dat	ta	n	desu
DM	relatively	morning	early	time	Copula	PAST	COMP	Copula(POL)

ga	ushi	no	naki	goe	ga	kikoe	te	eh	kumade
but	cow	GEN	bark	voice	SUBJ	hear	and	DM	rake

de	bokuso	o	kaki	wakeru	oji	san	to	oba	san
with	grass	ACC	rake	aside	uncle	SUFFIX	and	aunt	SUFFIX

ga	mi	e	te	eh	burasagat	teiru	hurui	toranjisuta
SUBJ	see	can	and	DM	hang	PROG	old	transistor

rajio	kara	desu		nee	hijouni	kou	tomeina	oto	de
radio	from	Copula(POL)		PART	very	DM	clear	sound	in

enuechike	anaunsa	no	koe	ga	kiko	e	mashi
NHK	announcer	GEN	voice	SUBJ	hear	can	Copula(POL)

ta	de	eh	sore	wa	desu		ne	boku	ga	izen
PAST	and	DM	that	TOP	Copula(POL)		PART	I	SUBJ	before

ano	nagano	ken		no	eh	kogen	noka	kogen
DM	Nagano	prefecture		GEN	DM	high plain	farmer	high plain

yasai	noka	de	hatara	itei	ta	toki	ni	desu
vegetable	farmer	at	work	PROG	PAST	time	at	Copula(POL)

ne ...
PART

"Well, well, in my close neighborhood, there is a cow stable, and eh, when I chose this apartment, one of the important decision-making factors was this stable. Why? Eh, when (I) passed by the stable, relatively early in the morning, (I heard) cow barks, and eh, (I) could see a middle-aged man and woman who were raking grass, and eh (I) could hear, say, very clear voice of the NHK announcer through the old transistor radio hanging (there). And that was, when I worked at a farm in Nagano Prefecture, ah, a farm at the high plain, a vegetable farm at the high plain. . . ."

References

Abutalebi, Jubin, Simona M. Brambati, J.-M. Annoi, Andrea Moro, Stefano F. Cappa, and Daniela Perani. (2007). The neural cost of the auditory perception of language switches: An event-related functional magnetic resonance imaging study in bilinguals. *The Journal of Neuroscience* 27(50), 13762–13769. https://doi.org/10.1523/NEUROSCI.3294-07.2007.

Abutalebi, Jubin, Pasquale A. Della Rosa, Guosheng Ding, Brendan Weekes, Albert Costa, and David W. Green. (2013). Language proficiency modulates the engagement of cognitive control areas in multilinguals. *Cortex* 49, 905–911. http://dx.doi.org/10.1016/cortex.2012.08.018.

Akinaga, Kazue (ed.). (2014). *Shin-meikai Nihongo Akusento Jiten* [Meikai Dictionary of Japanese Accent] Version 2. Tokyo: Sanseido.

Baumann, Stefan, and Arndt Riester. (2013). Coreference, lexical givenness, and prosody in German. *Lingua* 136, 16–37. http://dx.doi.org/10.1016/j.lingua.2013.07.012.

Baumann, Stefan, and Petra B. Schumacher. (2020). The incremental processing of focus, givenness, and prosodic prominence. *Glossa* 5(1), 6, 1–30. https://doi.org/10.5334/gjgl.914.

Baumann, Stephan, and Bodo Winter. (2018). What makes a word prominent?: Predicting untrained German listeners' perceptual judgment. *Journal of Phonetics* 70, 20–38. https://doi.org/10.1016/j.wocn.2018.05.004.

Belio, Julie, and Anne Lacheret. (2015). Disfluency and discursive markers: When prosody and syntax plan discourse. *Proceedings of Disfluency in Spontaneous Speech (DiSS) 2013*, 5–7. http://www.isca-speech.org/archive.

Bishop, Jason B. (2011). English listeners' knowledge of the broad versus narrow focus contrast. *Proceedings of International Society of Phonetic Sciences (ICPhS)* XVII, 312–315.

Bishop, Jason B., Grace Kuo, and Boram Kim. (2020). Phonology, phonetics, and signal-extrinsic factors in the perception of prosodic prominence: Evidence from

rapid prosody transcription. *Journal of Phonetics* 82, 1–20. https://doi.org/10.1016
/j.wocn.2020.100977.

Boersma, Paul, and David Weenink. (2018). Praat: Doing phonetics by computer. Version 6.1.53. Retrieved September 9, 2021, from http://www.praat.org/.

Bolinger, Dwight. (1961a). Contrastive accent and contrastive stress. *Language* 37, 83–96.

Bolinger, Dwight. (1961b). Ambiguities in pitch accent. *Word* 17, 309–317.

Bolinger, Dwight. (1972). *That's That*. The Hague: Mouton.

Bolinger, Dwight. (1986). *Intonation and Its Uses: Melody and Spoken English*. Stanford: Stanford University Press.

Branzi, Francesca M., Pasquale A. Della Rosa, Matteo Canini, Albert Costa, and Jubin Abutalebi. (2015). Language control in bilinguals: Monitoring and response selection. *Cerebral Cortex* 2015, 1–14. https://doi.org/10.1093/cercor/bhv052.

Breen, Mara, Edward Gibson, and Michael Wagner. (2010). Acoustic correlates of information structure. *Language and Cognitive Processes* 25(7), 1044–1098. http://dx.doi.org/10.1080/01690965.2010.504378.

Büring, Daniel. (2015). Unalternative semantics. *SALT* 25, 550–575. https://doi.org/10.3765/salt.v25i0.3634.

Büring, Daniel. (2016). *Intonation and Meaning*. Oxford: Oxford University Press.

Calhorn, Sacha. (2010). The centrality of metrical structure in signaling information structure: A probabilistic perspective. *Language* 86(1), 1–42.

Cangemi, Francesco, and Stefan Baumann. (2020). Integrating phonetics and phonology in the study of linguistic prominence. *Journal of Phonetics* 81, 1–6. https://doi.org/10.1016/j.wocsn.2020.100993.

Chafe, Wallace. (1970). *The Meaning and the Structure of Language*. Chicago: Chicago University Press.

Chien, Pei-Ju, Angela D. Friederici, Gesa Hartwigsen, and Daniela Sammler. (2020). Neural correlates of intonation and lexical tone in tonal and non-tonal language speakers. *Human Brain Mapping* 41(7), 1–17. https://doi.org/10.1002/hbm.24916.

Choi, Hye-Won. (1997). Topic and focus in Korean: The information partition by phrase structure and morphology. *Japanese/Korean Linguistics* 6, 545–561.

Chomsky, Noam, and Morris Halle. (1968). *The Sound Pattern of English*. New York: Harper and Row.

Christoffels, Ingrid, K., Elia Formisano, and Niels O. Schiller. (2007). Neural correlatates of verbal feedback processing: An fMRI study employing overt speech. *Human Brain Mapping* 28, 868–879. https://doi.org/10.1002.hbm.20315.

Cinque, Guglielmo. (1993). A null theory of phrase and compound stress. *Linguistic Inquiry* 24(2), 239–297.

Cole, Jennifer, José I. Hualde, Caroline L. Smith, Christopher Eager, Timothy Mahrt, and Ricardo Napoleão de Souza. (2019). Sound, structure and meaning: The bases of prominence ratings in English, French and Spanish. *Journal of Phonetics* 75, 113–147. https://doi.org/10.1016/j.wocn.2019.05.002.

Cole, Jennifer, Yun-S. Mo, and S.-D. Beak. (2010a). The role of syntactic structure in guiding prosody perception with ordinary listeners and everyday speech. *Language and Cognitive Processes* 25, 1141–1177. https://doi.org/10.1080/01690960903525507.

Cole, Jennifer, and Stefanie Shattuck-Hufnagel. (2016). New methods for prosodic transcription: Capturing variability as a source of information. *Laboratory Phonology: Journal of the Association for Laboratory Phonology* 7(1), 1–29. http://dx.doi.org/10.5334/labphon.29.

Cole, Jennifer, Yun-S. Mo, and Mark Hasegawa-Johnson. (2010b). Signal-based and expectation-based factors in the perception of prosodic prominence. *Laboratory Phonology* 1, 425–452. https://doi.org/10.1515/labphon.2010.022.

Cutler, Ann, and Takashi Otake. (1999). Pitch accent in spoken-word recognition in Japanese. *Journal of the Acoustical Society of America* 105(3), 1877–1888. https://doi.org/10.1121/1.426724.

Dehaene, Stanislas, Emmanuel Dupoux, Jacques Mehler, Laurent Cohen, Eraldo Paulesu, Daniela Perani, Pierre-Francois van de Moortele, Stéphane Lehéciry, and Denis Le Bihan. (1997). Anatomical variability in the cortical representation of first and second languages. *NeuroReport* 8, 3809–3815.

Féry, Caroline. (2017). *Intonation and Prosodic Structure*. Cambridge: Cambridge University Press.

Fukui, Naoki (1995). *Theory of Projection in Syntax*. Tokyo: Kurosio Publishers.

Gaab, Nadine, Christian Gaser, Tino Azehle, Lutz Jancke, and Gottfried Schlaug. (2003). Functional anatomy of pitch memory: An fMRI study with sparse temporal sampling. *NeuroImage* 19, 1417–1426. https://doi.org/10.1016/S1053-8119(03) 00225-6.

Gandour, Jackson. (1983). Tone perception in far Eastern languages. *Journal of Phonetics* 11, 149–175. https://doi.org/10.1016/S0095-4470(19)30813-7.

Gandour, Jackson, Donald Wong, Li Hsieh, Bret Weinzaphel, Diana Van Lancker, and Gary D Hutchins. (2000). A crosslinguistic PET study of tone perception. *Journal of Cognitive Neuroscience* 12(1), 207–222. https://doi.org/10.1162/089892900561841.

Gandour, Jackson, Yunxia Tong, Thomas Talavage, Donald Wong, Mario Dzemidzic, Yisheng Xu, Xiaojian Li, and Mark Lowe. (2007). Neural basis of first and second language processing of sentence-level linguistic prosody. *Human Brain Mapping* 28, 94–108. https://doi.org/10.1002/hbm.20255.

Geiser, Eveline, Tino Zaehle, Lutz Laucke, and Martin Meyer. (2008). The neural correlate of speech rhythm as evidenced by metrical speech processing. *Journal of Cognitive Neuroscience* 20(3), 541–552. https://doi.org/10.1162/jocn.2008. 20029.

Green, David, and Jubin Abutalebi. (2013). Language control in bilinguals: The adaptive control hypothesis. *Studies of Cognitive Psychology* 25(5), 515–530. https://doi.org/10.1080/20445911.2013.796377.

Grice, Martine, D. Robert Ladd, and Amalia Araniti. (2000). On the place of phrase accents in intonational phonology. *Phonology* 17, 143–185. https://doi.org/10.1017/S0952675700003924.

Gussenhoven, Carlos. (2007). Types of focus in English. In Chungmin Lee, Matthew Gordon, and Daniel Büring (Eds.), *Topic and Focus*, 83–100. Dordrecht: Springer.

Gussenhoven, Carlos. (2015). Does phonological prominence exist? *Lingue e Linguaggio* XIV(1), 7–24. https://doi.org/10.1419/80751.

Gussenhoven, Carlos, B. H. Repp, A. Rietveld, H. H. Rump, and Jacques Terken. (1997). The perceptual prominence of fundamental frequency peaks. *The Journal of the Acoustical Society of America* 102(5 Pt1), 3009–3022. https://doi.org/10.1121/1.420355.

Halliday, M. A. K. (1967–1968). Notes on transitivity and theme in English Part 1 – Part 3. *Journal of Linguistics* 3(1), 37–81, *Journal of Linguistics* 3(2), 199–244, *Journal of Linguistics* 4(2), 179–215. https://doi.org/10.1017/S002226700016613 (Part 2), https://doi.org/10.1017/S0022226700001882 (Part 3).

Hara, Yurie. (2005). Contrastives and Gricean principle. In *Proceedings of the 15th Amsterdam Colloquium*, 101–106.

Hara, Yurie. (2006). *Grammar of knowledge representation: Japanese discourse items at interfaces*. Doctoral Dissertation, University of Delaware.

Hesling, Isabelle, Bixente Dilharreguay, Martine Bordessoules, and Michèl Allard. (2012). The neural processing of second language comprehension modulated by the degree of proficiency: A listening connected speech fMRI study. *The Open Neuroimageing Journal* 4, 44–54. https://doi.org/10.2174/1874440001206010044.

Heycock, Caroline. (1994). Focus projection in Japanese. *North East Linguistics Society* 24(1), Article 12. https://scholarworks.umass.edu/nels/vol24/iss1/12.

Heycock, Caroline. (2008). Japanese *-wa*, *-ga*, and information structure. In Miyagawa and Saito (Eds.), 2008, 54–83.

Hickok, Gregory, and David Poeppel. (2000). Towards a functional neuroanatomy of speech perception. *Trends in Cognitive Sciences* 4, 131–138. https://doi.org/10.1016/s1364-6613(00)01463-7.

Howell, Jonathan, Mats Rooth, and Michael Wagner. (2017). Acoustic classification of focus: On the web and in the lab. *Laboratory Phonology: Journal of the Association for Laboratory Phonology* 8(1), 16. http://doi.org/10.5334/labphon.8.

Hu, Xiaochen, Hermann Ackermann, Jason A. Martin, Michael Erb, Susanne Winkler, and Suzanne M. Reiterer. (2013). Language aptitude of pronunciation in advanced second language (L2) learners: Behavioral predictors and neural substrates. *Brain and Language* 127(3), 366–376. http://dx.doi.org/10.1016/j.bandl.2012.11.006.

Huang, Tsan, and Keith Johnson. (2010). Language specificity in speech perception: Perception of Mandarin tones by native and nonnative listeners. *Phonetica* 67, 243–267. https://doi.org/10.1159/000327392.

Igarashi, Yosuke, and Hanae Koiso. (2012). Pitch range control of Japanese boundary pitch movements. *InterSpeech* 2012, 1949–1952.

Igarashi, Yusuke. (2014). Typology of prosodic phrasing in Japanese dialects. In Jun (Ed.), 464–492.

Ischebeck, Anja K., Angela D. Friederici, and Kai Alter. (2008). Processing prosodic boundaries in natural and hummed speech: An fMRI study. *Cerebral Cortex* 18, 541–522. https://doi.org//10.1093/cercor/bhm083.

Ishihara, Shin'ichiro. (2003). *Intonation and interface conditions*. Doctoral Dissertation, MIT.

Ishihara, Shin'ichiro. (2011). Japanese focus prosody revisited: Freeing focus from prosodic phrasing. *Lingua* 121(13), 1870–1889. https://doi.org/10.1016/j.lingua.2011.06.008.

Ishihara, Shin'ichiro. (2016). Japanese downstep revisited. *Natural Language and Linguistic Theory* 34, 1389–1443. https://doi.org/10.1007/s11049-015-9322-8.

Ishihara, Shin'ichiro, Yoshihisa Kitagawa, Satoshi Nambu, and Hajime Ono. (2018). Non-focal prominence. In C. Gulliemot, T. Yoshida, and S. Lee (Eds.), *Proceedings of the 13th Workshop on Altaic Formal Linguistics* (*WAFL* 13, *MIT Working Papers in Linguistics*, Vol. 88), 243–254. MITWPL.

Ito, Junko, and Armin Mester. (2012). Recursive prosodic phrasing in Japanese. In Toni Borowsky, Shigeto Kawahara, Takahiro Shinya, and Mariko Sugahara (Eds.), *Prosody Matters*, 280–303. London: Equinox.

Ito, Junko, and Armin Mester. (2013). Prosodic subcategories in Japanese. *Lingua* 124, 20–40. https://doi.org/10.1016/j.lingua.2012.08.016.

Jackendoff, Ray. (1972). *The Semantic Interpretation in Generative Grammar*. Cambridge, MA: MIT Press.

Jun, Sun-Ah. (Ed.). (2007). *Prosodic Typology: The Phonology of Intonation and Phrasing*. Oxford: Oxford University Press.

Jun, Sun-Ah. (Ed.). (2014). *Prosodic Typology II*. Oxford: Oxford University Press.

Jyothi, Preethi, Jennifer Cole, Mark Hasegawa-Johnson, and Vandana Puri. (2014). An investigation of prosody in Hindi narrative speech. *Proceedings of Speech Prosody* 7, 623–627. https://doi.org/10.21437/SpeechProsody.

Kadmon, Nirit, and Aldo Sevi. (2011). Without focus. *Baltic International Yearbook of Cognition, Logic and Communication* 6(1), 1–50. https://doi.org/10.4148/biyclc.v6i0.1585.

Kawakami, Shin. (1957). Tokyogo-no takuritu kyocho-no oncho [Prominent emphatic prosody of Tokyo Japanese]. *Reprinted in Kawakami 1995*, 76–91.

Kawakami, Shin. (1963). Bunmatsu-nadono joshocho-ni tsuite [On the rising tone at sentence-final and so on]. *Reprinted in Kawakami 1995*, 274–198.

Kawakami, Shin. (1995). *Nihongo Akusento Ronshu* [Anthology on Japanese Accent]. Tokyo: Kyuko Shoin.

Kayne, Richard. (1994). *The Antisymmetry of Syntax*. Cambridge, MA: MIT Press.

Kiss, Katalin É. (1998). Identificational focus and information focus. *Language* 74(2), 245–273. https://doi.org/10.2307/417867.

Kitahara, Mafuyu. (2001). *Category structure and function of pitch accent in Tokyo Japanese*. Doctoral Dissertation, Indiana University.

Kitahara, Mahuyu, and Shigeaki Amano. (2001). Perception of pitch accent categories in Tokyo Japanese. *Gengo Kenkyu* 120, 1–34. https://doi.org/10.11435/gengo1939.2001.120_1.

Klok, Rozina V., Heather Goad, and Michael Wagner. (2018). Prosodic focus in English vs. French: A scope account. *Glossa* 3(1), 71, 1–47. https://doi.org/10.5334/gjgl.172.

Koiso, Hanae. (Ed.). (2015). *Hanashi Kotoba Kopasu: Sekkei to Kochiku* [Corpus of Spontaneous Speech: Design and Construction]. Tokyo: Asakura Publishers.

Kori, Shiro. (1989). Kyocho-to intoneshon [Emphasis and intonation]. In Miyoko Sugito (Ed.), *Nihongo-no Onsei On'in Part I* [Japanese Phonetics and Phonology], 316–342. Tokyo: Meiji Shoin.

Kori, Shiro. (2018). Kanto-joshi-no intoneshon-to kanto-joshi-teki intoneshon [Intonation of Japanese interjection particles and sentence-internal interjectional intonation: Exemplification of the use with speech data]. *Gengo Bunka Kenkyu* 44, 283–306. https://doi.org/10.18910/68025.

Kori, Shiro. (2020). *Nihongo-no Intoneshon* [Intonation of Japanese]. Tokyo: Taishukan.

Kotz, Sonija, and Michael Schwartze. (2010). Cortical speech processing unplugged: A timely subcortico-cortical framework. *Trends in Cognitive Sciences* 14(9), 392–399. https://doi.org/10.1016/j.tics.2010.06.005.

Kratzer, Angelika, and Elizabeth Selkirk. (2020). Deconstructing information structure. *Glossa* 5(1), 1–53. https://doi.org/10.5334/gjgl.968.

Krifka, Manfred. (2008). Basic notions of information structure. *Acta Linguistica Hungarica* 55(3–4), 243–276. https://doi.org/10.1556/ALing.55.2008.3-4.2.

Kubozono, Haruo. (1988). *The organization of Japanese prosody*. Doctoral Dissertation, University of Edinburgh.

Kubozono, Haruo. (2007). Focus and intonation in Japanese: Does focus trigger pitch reset? *Workshop on Prosody Syntax, and Information Structure* (WPSI) 2, 1–27.

Kuno, Susumu. (1973). *The Structure of the Japanese Language*. Cambridge, MA: MIT Press.

Kuroda, S.-Y. (1965). *Generative Grammatical Studies in the Japanese Language*. New York: Garland Publishers.

Kuroda, S.-Y. (2013). Prosody and the syntax of indeterminates. *Lingua* 124, 64–95. http://dx.doi.org/10.1016/j.lingua.2012.10.013.

Ladd, D. R. (1996). *Intonational Phonology*. Cambridge: Cambridge University Press.

Ladd, D. R., Jo Verhoeven, and Karen Jacobst. (1994). Influence of adjacent pitch accents on each other's perceived prominence: Two contradictory effects. *Journal of Phonetics* 22(1), 87–99. https://doi.org/10.1016/S0095-4470(19)30268-2.

Lai, Yuwen. (2008). *Acoustic realization and perception of English lexical stress by Mandarin learners*. Doctoral Dissertation, University of Kansas.

Lee, Yong-Cheol, Bei Wang, Sisi Chen, Martine Adda-Decker, Angelique Amelot, Satoshi Nambu, and Mark Liberman. (2015). A cross-linguistic study of prosodic focus. In *IEEE International Conference on Acoustics, Speech and Signal Processing (ICASSP)*, 4754–4758. https://doi.org/10.1109/ICASSP.2015.7178873.

Lee, Y.-C., Satoshi Nambu, and Sunghye Cho. (2019). Dataset of focus prosody in Japanese phone numbers. *Data in Brief* 25. https://doi.org/10.1016/j.dib.2019.104139.

Liberman, Mark. (1975). *The intonational system of English*. Doctoral Dissertation, MIT.

Lœvenbruck, H., R. Grandchamp, L. Rapin, J. Perrone-Bertolotti, C. Pichat, C. Haldin, E. Cousin, J. P. Lachaux, M. Dohen, M. Perrier, and J. Baciu. (2019). Neural correlates of inner speech speaking, imitating and hearing: An fMRI study.

In *Proceedings of the 19th International Congress of Phonetic Sciences (ICPhS)* 2019, 1407–1411.

Luchkina, Tatiana, and Jennifer Cole. (2014). Structural and prosodic correlates of prominence in free word order language discourse. *Proceedings of Speech Prosody* 7, 1119–1123. https://doi.org/10.21437/SpeechProsody.

Luchkina, Tatiana, and Jennifer S. Cole. (2019). Perception of word-level prominence in free word order language discourse. *Language and Speech*, 1–32. https://doi.org /10.1177/0023830919884089.

Ma, Hengfen, Jiehui Hu, Jie Xi, Wen Shen, Jianqiao Ge, Feng Gen, Yuntao Wu, Jinjin Guo, and Dezhong Yao. (2014). Bilingual cognitive control in language switching: An fMRI study of English-Chinese late bilinguals. *PLoS ONE* 9(9), e106468. https://doi.org/10.1371/journal.pone.016468.

Maeda, Kazuaki, and Jennifer J. Venditti. (1998). Phonetic investigation of boundary pitch movements in Japanese. In *Proceedings of the 5th International Conference on Spoken Language Processing (ICSLP 98)*. http://www.isca-speech.org/archive.

Maekawa, Kikuo. (2011a). *Kopasu-o riyo-shita shizen-onsei-no kenkyu* [Corpus-based study on natural speech]. Doctoral Dissertation, Tokyo Institute of Technology.

Maekawa, Kikuo. (2011b). PNLP-no onsei-teki keijo-to gengo-teki kino [Phonetic shape and linguistic function of penultimate non-lexical prominence]. *Journal of the Phonetic Society of Japan* 15(1), 16–28.

Mahrt, Timothy. (2016). LMEDS: Language makeup and experimental design software. https://github.com/timmahrt/LMEDS.

Marcus, M. P., B. Santorini, and M. A. Marcinkiewicz. (1993). Building a large annotated corpus of English: The Penn Treebank. *Computational Linguistics* 19, 313–330.

McCawley, James. (1968). *The Phonological Component of a Grammar of Japanese.* The Hague: Mouton.

Meyer, Martin, Kai Alter, Angela Friederici, Gabriele Lohmann, and D. Yves von Cramon. (2002). FMRI reveals brain regions mediating slow prosodic modulations in spoken sentence. *Human Brain Mapping* 17, 73–88. https://doi.org/10.1002/ hbm.10042.

Meyer, Martin, Karsten Steinhauer, Kai Alterm, Angela D. Friederici, and D. Yves von Cramon. (2004). Brain activity varies with modulating of dynamic pitch variance in sentence melody. *Brain and Language* 89, 277–289. https://doi.org/10 .1016/S0093-934X(03)00350-X.

Milsark, Gary L. (1979). *Existential sentences in English.* Doctoral Dissertation, MIT.

Miyagawa, Shigeru, and Mamoru Saito (Eds.). (2008). *The Oxford Handbook of Japanese Linguistics.* Oxford: Oxford University Press.

Mizuguchi, Shinobu, Gabor Pintér, and Koichi Tateishi. (2017). Natural speech perception cues by Japanese learners of English. *Proceedings of Pacific Second Language Research Forum (PacSLRF)* 2016, 151–156.

Mizuguchi, Shinobu, and Koichi Tateishi. (2018). Focus prosody in Japanese reconsidered. *2018 Proceedings of the Linguistic Society of America (PLSA)* 3(28), 1–15. https://doi.org/10.3765/plsa.v3il.4291.

Mizuguchi, Shinobu, Tomothy Mahrt, and Koichi Tateishi. (2019a). How L2 learners perceive English prosody. *Proceedings of the 2nd International Symposium on Applied Phonetics*. https://www.isca-speech.org/archive/ISAPh_2018/.

Mizuguchi, Shinobu, Yukiko Nota, and Koichi Tateishi. (2019b). Perception of narrow focus by bilingual speakers. *Proceedings of the 19th International Congress of Phonetic Sciences (ICPhS)* 2019, 1660–1664.

Mizuguchi, Shinobu, and Koichi Tateishi. (2020a). Why is L1 not easy to hear. *2020 Proceedings of Linguistic Society of America (PLSA)* 5(1), 59–73. https://doi.org/10.3765/plsa.v5i1.4668.

Mizuguchi, Shinobu, and Koichi Tateishi. (2020b). Prominence in Japanese is not only cued acoustically. *Proceedings of Speech Prosody 2020*. https://doi.org/10.21437/SpeechProsody.2020-24.

Mizuguchi, Shinobu, and Koichi Tateishi. (2022). Perception of boundary and prominence in spontaneous Japanese: An RPT study. *Proceedings of Speech Prosody 2022*, 649–653. https://doi.org/10.21327/SpeechProsody.2022-132.

Nan, Yun, and Angela D. Friederici. (2012). Differential roles of right temporal cortex and broca's area in pitch processing: Evidence from music and Mandarin. *Human Brain Mapping* 34(9), 2045–2052. https://doi.org/10.1002/hbm.22046.

Nespor, Marina, and Irene Vogel. (1986). *Prosodic Phonology*. Dordrecht: Foris.

Oishi, Hatsutaro. (1959). Purominensu-ni tsuite [On prominence: Note on Tokyo Japanese]. *Kotoba-no Kenkyu* 1, 87–102. http://doi.org/10.15084/00001705.

Ortega-Llebaria, Marta, and Laura Colantoni. (2014). L2 English intonation: Relations between form-meaning associations, access to meaning and L1 transfer. *Studies in Second Language Acquisition* 36, 331–353. https://doi.org/10.1017/S0272263114000011.

Perrone-Bertolletti, Marcela, Marion Dohen, Helen Loevenbruck, Marc Sato, Cédric Pichat, and Monica Baciu. (2013). Neural correlates of the perception of contrastive prosodic focus in French: A functional magnetic resonance imaging study. *Human Brain Mapping* 34, 2574–2591. https://doi.org/10.1002/hbm.22090.

Pierrehumbert, Janet. (1979). The perception of fundamental frequency declination. *Journal of Acoustic Society of America* 66(2), 363–369. https://doi.org/10.1121/1.383670.

Pierrehumbert, Janet. (1980). *The phonology and phonetics of English intonation*. Doctoral Dissertation, MIT.

Pierrehumbert, Janet, and Mary Beckman. (1986). Intonational structure in Japanese and English. *Phonology* 3, 255–309. https://doi.org/10.1017/S09526757000006X.

Pierrehumbert, Janet, and Mary Beckman. (1988). *The Japanese Tone Structure*. Cambridge, MA: MIT Press.

Pintér, Gabor, Shinobu Mizuguchi, and Kazuhito Yamato. (2014a). Boundary and prominence perception by Japanese learners of English: A preliminary study. *Phonological Studies* 17, 59–66.

Pintér, Gabor, Shinobu Mizuguchi, and Koichi Tateishi. (2014b). Perception of prosodic prominence and boundaries by L1 and L2 speakers of English. *Proceedings of InterSpeech* 2014, 544–547. http://www.isca-speech.org/archive.

Pitt, Mark A., Laura Dilley, Keith Johnson, Scott Klesling, William Raymond, Elizabeth Hume, Scott Kiesling, and William Raymond. (2007). *Buckeye Corpus of Conversational Speech* (2nd Release). Columbus, OH: Department of Psychology, Ohio State University. www.buckeyecorpus.osu.edu.

Poser, Bill. (1984). *The phonetics and phonology of tone and intonation in Japanese.* Doctoral Dissertation, MIT.

Rietveld, A. C. M., and Carlos Gussenhoven. (1985). Perceived speech rate and intonation. *Journal of Phonetics* 15(3), 273–285. https://doi.org/10.1016/S0095-4470(19)30571-6.

Rochemont, Michael. (1986). *Focus in Generative Grammar.* Amsterdam: John Benjamins.

Rochemont, Michael. (2016). Givenness. In Caroline Féry and Shin'ichiro Ishihara (Eds.), *The Oxford Handbook of Information Structure*, 41–63. https://10.1093/oxfordhb/9780100642670.013.

Rooth, Mats. (1986). *Association with focus.* Doctoral Dissertation, University of Massachusetts at Amherst.

Rooth, Mats. (1992). A theory of focus interpretation. *Natural Language Semantics* 1, 75–116.

Rooth, Mats. (1996). Focus. In Shalom Lappin (Ed.), *The Handbook of Contemporary Semantic Theory*, 271–297. London: Blackwell.

Rooth, Mats. (2016). Alternative semantics. In Caroline Féry and Shin'ichiro Ishihara (Eds.), *The Oxford Handbook of Information Structure*, 260–288. Oxford: Oxford University Press.

Saito, Mamoru, and Naoki Fukui. (1998). Order in phrase structure and movement. *Linguistic Inquiry* 29(3), 439–447. https://doi.org/10.1162/002438998553815.

Sato, Marc, Pascale Tremblay, and Vincent L. Gracco. (2009). A mediating role of the premotor cortex in phoneme segmentation. *Brain and Language* 111(1), 1–9. https://doi.org/10.1016/j.bandl.2009.03.002.

Sawada, Osamu. (2007). The Japanese contrastive *wa*: A mirror image of *even. The Proceedings of the 33rd Annual Meeting of the Berkeley Lintuisitcs Society*, 374–386. https://doi.org/10.3765/bls.v33i1.3541.

Sawada, Osamu. (2019). Scalarity and alternatives of Japanese mora (letter)-based minimizers. *2019 Proceedings of Linguistic Society of America (PLSA)* 4, 1–15. https://doi.org/10.3765/plsa.v4i1.4525.

Schaal, Nora K., Marina Kretschmer, Ariane Keitel, Vanessa Krause, Jasmin Pfeifer, and Bettina Pollok. (2017). The significance of the right dorsolateral prefrontal cortex for pitch memory in non-musicians depends on baseline pitch memory abilities. *Frontiers in Neuroscience* 11, 1–9. https://doi.org/10.3389/fnins.2017.00677.

Schubotz, Ricarda I., Angela, D. Friederici, and D. Yves von Cramon. (2000). Time perception and motor timing: A common cortical and subcortical basis revealed by fMRI. *NeuroImage* 11, 1–12. https://doi.org/10.1006/nimag.1999.0514.

Schwarzschild, Roger. (1999). Givenness, avoid F and other constraints on the placement of focus. *Natural Language Semantics* 7(2), 141–177. https://doi.org/10.1023/A:1008370902407.

Selkirk, Elizabeth. (2009). On clause and intonational phrase in Japanese: The syntactic grounding of prosodic constituent structure. *Gengo Kenkyu* 136, 35–73.

Selkirk, Elizabeth, and Koichi Tateishi. (1988). Constraints on minor phrase formation in Japanese. In *Proceedings of the 24th annual meeting of the Chicago Linguistic Society*, 316–336.

Shinya, Takahiro. (1999). Eigo to Nihongo ni okeru fokasu ni yoru daunsuteppu no soshi to cho'on-undo no chogo [The blocking of downstep by focus and articulatory overlap in English and Japanese]. *Proceedings of Sophia University Linguistic Society* 14, 35–51.

Shinya, Takahiro. (2009). *The role of lexical contrast in the perception of intonational prominence in Japanese*. Doctoral Dissertation, University of Massachusetts at Amherst.

Shport, I. A. (2015). Perception of acoustic cues to Tokyo Japanese pitch-accent contrasts in native Japanese and native English listeners. *Journal of the Acoustical Society of America* 138(1), 307–318. http://doi.org/10.1121/1.4922468.

Skipper, Jeremy I., Howard C. Husbaum, and Steven L. Small. (2005). Listening to talking faces: Motor cortical activation during speech perception. *NeuroImage* 25, 76–89. https://doi.org/10.1016/j.neuroimage.2004.11.006.

Smith, Caroline. (2011). Perception of prominence and boundaries by naïve French listeners. *Proceedings of the 17th International Congress of Phonetic Sciences (ICPhS)*, 1874–1877.

Steffman, Jeremy. (2021). Prosodic prominence effects in the processing spectral cues. *Language, Cognition and Neuroscience* 36(5), 586–611. https://doi.org/10.1018/23273798.2020.1862259.

Sugahara, Mariko. (2002). *Downtrends and post-focus intonation in Tokyo Japanese*. Doctoral Dissertation, University of Massachusetts at Amherst.

Sugito, Miyoko. (1980). Thoughts on peak delay: An instrumental study of Japanese accent [in Japanese]. In Munemasa Tokugawa (Ed.), *Akusento [Accent]*, 65–94. Tokyo: Yuseido.

Sugito, Miyoko. (2001). Bunpo-to Nihongo-no akusento oyobi intoneshon [Characteristics of word accent and intonation of Tokyo and Osaka Japanese with relation to grammar]. In Spoken Language Working Group (Ed.), *Speech and Grammar* 3, 197–210. Tokyo: Kurosio Publishers.

Swerts, Marc. (1998). Filled pauses as markers of discourse structure. *Journal of Pragmatics* 30, 485–496.

Taniguchi, Miki, and Takehiko Maruyama. (2001). Shoten kozo-to joshi-no takuritsu [Focus and prominence on particles in Japanese]. *Proceedings of Kansai Linguistic Society* 21, 56–66.

Tomioka, Satoshi. (2009). Contrastive topics operate on speech acts. In Malte Zimmermann and Caroline Féry (Eds.), *Information Structure: Theoretical, Typological, and Experimental Perspectives*, 115–138. Oxford: Oxford University Press.

Tonhauser, Judith. (2019). Prosody and meaning: On the production, perception and interpretation of prosodically realized focus. In Chris Cummins and Napoleon

Katsos (Eds.), *Handbook of Experimental Semantics and Pragmatics*, 494–511. Oxford: Oxford University Press.

Van der Burght, Constantijn L., Tomás Goucha, Angela D. Friederici, Jens Kreitewolf, and Gesa Hartwigsen. (2019). Intonation guides sentence processing in the left frontal gyrus. *Cortex* 117, 122–134. https://doi.org/10.1016/j.cort3ex. 2019.02.011.

Venditti, Jennifer J. (1997). Japanese ToBI labelling guidelines. *Ohio State University Working Papers in Linguistics* 50, 127–162. (First distributed in 1995).

Venditti, Jennifer J. (2007). The J_ToBI model of Japanese intonation. In Jun (Ed.), 172–200.

Venditti, Jennifer J., Kikuo Maekawa, and Mary Beckman. (2008). Prominence marking in the Japanese intonation system. In Miyagawa and Saito (Eds.). 2008, 456–512.

Wagner, Michal. (2020). Prosodic focus. In Daniel Gutzmann, Lisa Matthewson, Cécile Meier, Hotze Rullmann, and Thomas E. Zimmermann (Eds.), *The Wiley Blackwell Companion to Semantics*. Hoboken, NJ: Wiley-Blackwell. https://doi. org/10.1002/9781118788516.sem133.

Watanabe, Michiko. (2009). *Features and roles of filled pauses in speech communication*. Doctoral Dissertation, The University of Tokyo.

Weissman, D. H., A. S. Perkins, and M. G. Woldorff. (2008). Cognitive control in social situations: A role for the dorsolateral prefrontal cortex. *Neuroimage* 40(2), 955–962. https://doi.org/10.1016/j.neuroimage.2007.12.021.

Wells, John C. (2006). *English Intonation: An Introduction*. Cambridge: Cambridge University Press.

Whitman, John. (2001). Kayne 1994: P.143, FN.3. In Galina M. Alexandrova and Olga Arnaudova (Eds.), *The Minimalist Parameter*, 77–100. Amsterdam: John Benjamins.

Wightman, Collin W., Mari Ostendorf, and Patti J. Price. (1992). Segmental durations in the vicinity of prosodic phrase boundaries. *The Journal of the Acoustical Society of America* 91(3), 1707–1717. https://doi.org/10.1121/1.402450.

Yamada, Reiko, Y. Tohkura, and N. Kobayashi. (1992). Effect of word familiarity on nonnative phoneme perception: Identification of English /r/, /l/ and /w/ by native speakers of Japanese. In Allan James and Jonathan Leather (Eds.), *Second Language Speech*, 103–117. The Hague: Mouton de Gruyter.

Zimmermann, Malte, and Caroline Féry (Eds.). (2010). *Information Structure: Theoretical, Typological, and Experimental Perspectives*. Oxford: Oxford University Press.

Zubizarreta, Maria L. (1998). *Prosody, Focus, and Word Order (Linguistic Inquiry Monographs)*. Cambridge, MA: MIT Press.

Index

Page references for figures are italicized

accented (A-) words, 2, 18–19, 26, 28, 32–35, 71, 82, 90
accentual boost, 28, 33, 34, 46; normalization of, 34, 35
accentual phrase (AP), 2, 4, 49n5, 75, 77
Adaptive Control Hypothesis, 57, 59
agglutinative language, 1, 17, 20, 64
alignment prominence. *See* non-focal prominence
Alternative Semantics, 13, 23, 81
American English, 38, 42, 43, 49n4, 93, 95

Basal ganglia, 57
baseline, *33*, 17
Baumann, Stefan, 23, 65, 84
bilingual, late-onset, 7, 8, 51, 52
bi-tonal pitch, 2, 17
Bolinger, Dwight, 1, 11
boundary, 6, 8, 9, 16, 17, 19, 34, 62–64, 67–73, 76–78, 82–84; coding of, 9, 70, 72, 73; salience of, 62, 63
boundary chunking, 6, 62, 77–78
boundary pitch movement (BPM), 9, 77, 80, 84
b(oundary)-score, 9, 62–64, *66*, 67–68, 78

boundary tone, 3, 5, 7–8, 18, 34, 35, 44, 45, 49, 71–73, 76, 81, 82, 84; prominence of, 7–8; semantic and pragmatic role of, 8, 34
broad focus (BF), 4, 5, 7, 11, 13–15, 20, 24, 25, 27, 28, 31–33, 35, 37, 38, 51–59, 60n1, 79, 86, 89, 95–97
Broadman Area (BA), 55
Buckeye Corpus, 62
Büring, Daniel, 1, 5, 6, 11, 13, 61

case particle, 16, 20, 23, 65
catathesis. *See* downstep
Cole, Jennifer, 9, 62–65, 67, 79, 84
comment, 14, 20
content word, 5, 9, 11, 14–16, 21, 23, 37, 65, 70–72, *73*, 79, 81, 82
contrastive focus. *See* corrective focus
Corpus of Spontaneous Japanese (CSJ), 9, *19*, 64, 72, *73*, 74n1, 77, 78, 84, 99
corrective focus (CF), 5, 7, 8, 13–16, 20, 24, 32, 37–42, 46, 49n1, 51–59, 61, 80, 81, 93, 95, 96; cross-linguistic variation of, 37–38, 52–57; in English, 7, 42, 52–57; identification of, 7, 38, 40, 42, 46, 52, *54*; in

About the Authors

Shinobu Mizuguchi is professor emeritus of linguistics at Kobe University and the author of *Individuation in Numeral Classifier Languages* (2004).

Koichi Tateishi is professor of linguistics at Kobe College and the author of *The Syntax of 'Subjects'* (1994).